Workshop on Challenges and Opportunities of Underground Coal Gasification in India (UCG-2017)

IOP Conference Series: Earth and Environmental Science
Volume 76

New Delhi, India
13 – 14 February 2017

ISBN: 978-1-5108-4509-1
ISSN: 1755-1307

Printed from e-media with permission by:

Curran Associates, Inc.
57 Morehouse Lane
Red Hook, NY 12571

Some format issues inherent in the e-media version may also appear in this print version.

Printed by Curran Associates, Inc. (2017)

For permission requests, please contact the Institute of Physics at the address below.

Institute of Physics
Dirac House, Temple Back
Bristol BS1 6BE UK

Phone: 44 1 17 929 7481
Fax: 44 1 17 920 0979

techtracking@iop.org

Additional copies of this publication are available from:

Curran Associates, Inc.
57 Morehouse Lane
Red Hook, NY 12571 USA
Phone: 845-758-0400
Fax: 845-758-2633
Email: curran@proceedings.com
Web: www.proceedings.com

Table of contents

Volume 76

Workshop on Challenges and Opportunities of Underground Coal Gasification in India (UCG-2017)

13–14 February 2017, New Delhi, India

Accepted papers received: 10 July 2017
Published online: 24 July 2017

Preface

Papers

IOP Conf. Series: Earth and Environmental Science **76** (2017) 011001 doi:10.1088/1755-1315/76/1/011001

Foreword

Coal is pivotal to meeting India's energy needs and with major international climate agreements; it has become a paramount objective to use coal in a sustainable manner. Also, while overall coal reserves in India are extremely large (more than 308 billion tonnes and comprising of 7% global reserves), the recoverable reserves are just over 60 billion tonnes [1]. Innovative coal mining technologies thus have to play an important role in ensuring that the recovery and utilization of coal improves in light of energy security for the country. For instance, it is reported that highwall mining technology has resulted in recovery of more than 25 lakh tonne of locked-up coal from three opencast mines [2]. It is here that the role of underground coal gasification (UCG) technology also becomes relevant. The technology has the potential to serve many of India's energy objectives i.e. increase of indigenous natural gas supply, extraction of energy from deep seated coal, reduction of fugitive methane emissions due to conventional coal mining and reduction in atmospheric emissions during power generation. This route may also ease the future into a hydrogen economy.

In 1984, I had the privilege of chairing a steering group on UCG in India. Subsequently, several other organizations (ONGC, Coal India, CSIR Laboratories and IITs) have performed considerable work in this area. As a result, a significant repository of geological characteristics, economic feasibility and mathematical models have been established within the country. While considerable work has been done in this area, many challenges remain for the commercialization of this technology – something that can be achieved through greater industry-academia collaboration. More than 25 years ago, I wrote, "The engineering challenges in exploiting our huge reserves of unmineable coals in north Gujarat (estimated at more than 60 billion tons), at a depth of 1000 m and more, are enormous and we can be world leaders" [3]. The same stands true for today as well and I am hopeful that this target may be realized to pave way for a sustainable future. In fact, the conditions in India are far more compelling and propitious than anywhere in the world and we may well find a great opportunity to become global leaders in UCG at depth of 1000 m and more.

I am happy to note that scientists at the CSIR-Central Institute of Mining and Fuel Research have been engaged in a 12th Five Year Plan Project on Underground Coal Gasification. As part of this project and with an effort to reach out to industry personnel, they organized a two-day workshop on "Challenges and Opportunities of Underground Coal Gasification in India". The results of this workshop are expected to be well-documented in the IOP Conference Series: Earth and Environmental Science. I am happy to note that the papers are a mix of industrial learning, lab-level experimentation and modelling, along with rigorous policy analysis. I commend the organizers for this timely initiative and hope that this leads to positive changes for this technology.

Professor M M Sharma, FRS, FREng, FNA
Emeritus Professor of Eminence, Institute of Chemical Technology, Mumbai

References

[1] Miller B G 2011 *Clean Coal Engineering Technology* (Oxford: Butterworth-Heinemann)
[2] Porathur J L, Verma C P, Pal Roy P, Prakash A 2017 *Proc NexGen Technologies for Mining and Fuel Industries* (New Delhi: Allied Publishers) pp 653-658
[3] Sharma M M 1991 *J. Indian Inst. Sci.* **71** 457-473

IOP Conf. Series: Earth and Environmental Science **76** (2017) 011001 doi:10.1088/1755-1315/76/1/011001

Preface

Coal is the mainstay of India's energy sector and is likely to shape India's energy future. With the Paris Climate Agreement, coupled with the Government of India's thrust on clean energy sources, it is imperative for us to deploy clean coal technologies with lesser environmental impacts. It should be ensured that such sources are offered at the lowest possible cost to ensure energy affordability to the lower strata of the society. While India has abundant coal reserves, much of the coal remains unmineable due to complex geo-mining conditions, mine hazards and economic complexities. If we are to continue using coal sustainably, different variants of utilizing energy from coal also need to be identified. Underground Coal Gasification (UCG) is one such technology. While the Government of India has shown significant support for this technology, that coupled with enthusiasm from the Industry has not translated into commercial utilization.

With an aim to further the cause of UCG in India, we undertook the Project CoalGasUrja (Development of Underground Coal Gasification Technology in India) as a CSIR Network Project under the Twelfth Five Year Plan. The project – which is now on the verge of completion – has resulted in significant outcomes for UCG in India in terms of coal characterization, reactivity studies, thermo-fluid modelling, site selection criteria and cost estimation. The final and perhaps the most crucial part of the project is knowledge dissemination to the industries. With this objective, a two-day workshop on Challenges and Opportunities of Underground Coal Gasification in India (UCG-2017) was organized during 13-14 February, 2017 at Vigyan Bhawan, New Delhi. It is a matter of immense pleasure for us that IOP Conference Series: Earth and Environmental Sciences is publishing the papers presented at the workshop. I am hopeful that this volume serves as a bridge for the next generation of scientists and engineers to carry out the work that we have left incomplete. The papers in this volume are a combination of policy, technical and industrial field experience. I thank all the authors who have contributed to this volume.

I thank our Director, Dr. Pradeep K Singh, Chairman, Organizing Committee and Dr. Rajendra Singh, Head of the Research Group for their guidance and personal support. I also thank Dr. Debadutta Mohanty, Joint Organizing Secretary and other members of the organizing committee for their skilful planning of the workshop. We are grateful to ONGC, JCB India, SAIL, ONGC Energy Centre, SCCL and NLC India, for providing generous financial support for this event. Of course, thanks are also due to all the other companies and academic institutions who have nominated experts to attend this event. Thanks are also due to Udayan for his active involvement at various stages starting from planning the event to preparing the workshop proceedings. Last but not the least, I express my sincere thanks to researchers of our group Jaywardhan, Akanksha, Sujoy, Om Prakash, Bageswari, Suman, Achinta and Anjali.

Ajay K Singh
Organizing Secretary, UCG-2017
Nodal Scientist, Project CoalGasUrja, CSIR-CIMFR, Dhanbad

IOP Conf. Series: Earth and Environmental Science **76** (2017) 011001 doi:10.1088/1755-1315/76/1/011001

Organizing Institution

CSIR-Central Institute of Mining and Fuel Research

Platinum Sponsors

Oil and Natural Gas Corporation

JCB India

Gold Sponsors

ONGC Energy Centre

Steel Authority of India Ltd.

Silver Sponsors

The Singareni Collieries Co. Limited

NLC India

IOP Conf. Series: Earth and Environmental Science **76** (2017) 011001 doi:10.1088/1755-1315/76/1/011001

Organizing Committee

Chairman: Dr. Pradeep K Singh, Director, CSIR-CIMFR, Dhanbad

Organizing Secretary: Dr. Ajay K Singh, Nodal Scientist, Project CoalGasUrja, CSIR-CIMFR

Joint Organizing Secretary: Dr. Debadutta Mohanty, Sr. Scientist, CSIR-CIMFR

Advisors:

- Prof. Durga C Panigrahi, Director, IIT (ISM), Dhanbad
- Prof. Harish Hirani, Director, CSIR-CMERI
- Dr. Virendra M Tiwari, Director, CSIR-NGRI
- Dr. Amalendu Sinha, Former Director, CSIR-CIMFR (Chair, Review Committee)
- Dr. Pijush Pal Roy, Outstanding Scientist, CSIR-CIMFR
- Prof. T Sundararajan, Professor, IIT Madras
- Prof. Amit Garg, Professor, IIM Ahmedabad
- Prof. Sanjay M Mahajani, Professor, IIT Bombay
- Prof. Anand B Rao, Associate Professor, IIT Bombay

Members:

- Dr. Kalendra B Singh, Head (NREM), CSIR-CIMFR
- Dr. Rajendra Singh, Head (SMNG), CSIR-CIMFR
- Prof. Atul K Varma, Professor, IIT (ISM), Dhanbad
- Dr H V S Satyanarayana, Sr. Pr. Scientist, CSIR-NGRI
- Dr. Harendra Singh, Sr. Pr. Scientist, CSIR-CIMFR
- Dr. Vinod A Mendhe, Pr. Scientist, CSIR-CIMFR
- Shri John Kispotta, Sr. Scientist, CSIR-CIMFR
- Dr. Dipankar Chatterjee, Sr. Scientist, CSIR-CMERI
- Shri Jaywardhan Kumar, Scientist, CSIR-CIMFR

Peer review statement

All papers published in this volume of *IOP Conference Series: Earth and Environmental Science* have been peer reviewed through processes administered by the proceedings Editors. Reviews were conducted by expert referees to the professional and scientific standards expected of a proceedings journal published by IOP Publishing.

IOP Conf. Series: Earth and Environmental Science **76** (2017) 012001 doi:10.1088/1755-1315/76/1/012001

Relevance of Clean Coal Technology for India's Energy Security: A Policy Perspective

Amit Garg[1,] *, Vineet Tiwari[2], Saritha Vishwanathan[1]

[1] Public Systems Group, Indian Institute of Management Ahmedabad, Vastrapur, Ahmedabad, India

[2] Department of Management Studies, Indian Institute of Information Technology – Allahabad, Allahabad, India

* Corresponding Author Email: amitgarg@iima.ac.in

Abstract. Climate change mitigation regimes are expected to impose constraints on the future use of fossil fuels in order to reduce greenhouse gas (GHG) emissions. In 2015, 41% of total final energy consumption and 64% of power generation in India came from coal. Although almost a sixth of the total coal based thermal power generation is now super critical pulverized coal technology, the average CO_2 emissions from the Indian power sector are 0.82 kg-CO_2/kWh, mainly driven by coal. India has large domestic coal reserves which give it adequate energy security. There is a need to find options that allow the continued use of coal while considering the need for GHG mitigation. This paper explores options of linking GHG emission mitigation and energy security from 2000 to 2050 using the AIM/Enduse model under Business-as-Usual scenario. Our simulation analysis suggests that advanced clean coal technologies options could provide promising solutions for reducing CO2 emissions by improving energy efficiencies. This paper concludes that integrating climate change security and energy security for India is possible with a large scale deployment of advanced coal combustion technologies in Indian energy systems along with other measures.

1. Introduction

India is an emerging economy with an average GDP growth of 6.7% over the last ten years and a projected rate of 7-7.3% for 2015-16. This is expected to significantly increase the demand for energy as per-capita energy consumption (PEC) (the ratio of the estimate of total energy consumption during the year to the estimated mid-year population of that year) increased from 13695 Mega Joules (MJ) in 2005-06 to 22725MJ in 2014-15, a CAGR of 5.59% [1]. The energy elasticity of the Indian economy has traditionally been above unity due to relatively lower energy efficiency in many energy intensive sectors and high coal dependence. To maintain its economic growth, India needs forward-looking energy efficiency enhancement policies combined with an integrated development of the energy sector. In view of climate change concerns, energy and technology choices in developed countries are being determined based on their capabilities to mitigate greenhouse gases (GHG) emissions. However, in India, the major focus is still on fulfilling energy demand while addressing local pollution concerns. Climate change security through lower GHG emissions and energy security due to higher energy availability need not be competing policy goals. Various clean coal technologies could provide an integration of these two policy goals. This paper tries to explore this through a bottom-up modelling analysis.

Coal being the dominant player of energy mix in India has various reasons for much wider acceptance. Its diverse geological distribution, abundant availability, matured industrial base, available technical competence makes it an obvious contender for leading the list of energy resources. Growing demand for energy and power in rapidly growing economy of the country is theoretically believed to be achieved with sustained domestic sources especially in a highly uncertain and fluctuating international energy markets. Suitability and safety of nuclear projects have been questioned in view of Fukushima accident in Japan. Renewable energy production options are unlikely to assume more than 20-30% of share in total energy supply in various estimated visions. Hydro potential is limited by the fact that country has limited perennial water bodies and most of them cater to the basic purpose of drinking, agriculture and other domestic uses. Therefore, hydro capacities alone cannot serve the growing demand. Due to very short switching time hydro power is preferable for meeting peak demands. Other fossils fuels like natural gas and oil are scarce and have very uncertain availability, pricing and trade treaties. Further these fuels have other important uses in domestic and transport sectors. Hence, coal is an obvious winner and leader in utility fuels in the country and likely to remain in future. But future concerns are directed towards greening coal in order to achieve sustainable development. Thus, coal-based clean technologies can be a sustainable choice to be adopted in Indian energy systems to meet growing energy needs.

In 2005, 72% of the CO_2 emissions in India came from coal, with a significant contribution from power and steel plants. The top three sectoral contributors were power plants (51.9%), steel plants (8.4%) and cement plants (8%) [2]. Many researchers [3,4,5,6] have analysed the contribution from various sectors. Some researchers have also analysed the contribution of different industrial plants. Garg and Shukla (2009) [2] indicated that the 25 largest emitters contributed 34% of the total all-India CO_2 emissions in 2005 and this number has grown by around 10% per annum during 1990–2005. This analysis shows that emissions are getting increasingly concentrated in a few major industrial units called large point sources (LPS), which provide focused mitigation opportunities.

Various international climate studies have indicated that the carbon dioxide (CO_2) proportion in the atmosphere has increased significantly due to heavy industrial activity during the last century. In 2013, global CO_2 emissions due to fossil fuel use (and cement production) were 36 gigatonnes (Gt CO_2); this is 61% higher than 1990 (the Kyoto Protocol reference year) and 2.3% higher than 2012 [7]. However, the repercussions of this increasing concentration have become apparent only recently. The major contributors to emissions which have adversely impacted the environment need to be identified and a suitable mechanism to tackle them has to be developed. Globally, the talks about finding solutions for mitigating energy related emissions have mainly emphasised three aspects, 1) improving efficiencies of technical systems, 2) promoting renewable energy sources in mainstream choices and 3) discouraging fossil use through regulation and taxation. Due to the prevailing energy dynamics in India, the use of fossil fuels, especially coal, is inevitable. Sustainable coal use can be ensured in three possible ways, 1) conditioning of coal before use, 2) efficient technology choices and 3) post combustion carbon capture.

The next section gives an overview of the coal use in India. In the third section, we provide the methodology and setting-up of the model for coal sector analysis and GHG emission mitigation possibilities under BAU scenario. The fourth section discusses the results followed by conclusions.

2. Coal demand and supply in India

The Indian coal industry is the fourth largest in terms of reserves and third largest by production in the world. Coal is accounting for about 70 per cent of the total energy consumption in the country. Coal deposits in India occur mostly in thick seams and at shallow depths. Indian coal has high ash content (15-45%) and low calorific value. With the present rate of daily coal extraction in the country, the reserves are likely to last over 100 years. India's coal production has increased to 612.44 million tonnes in 2015, more than the 565.77 million tonnes the country produced in the previous year.

As per the report of the Working Group on Coal [8], the demand for coal is likely to continue in the long term due to its dominance in the evolving energy mix, given the domestic availability of coal and improvement in the plant load factor of existing plants. The coal requirement has been projected at 980.5 Mt for 2016-17 implying a CAGR of 7.1% during 2012-2017, with 738 Mt for power, 67.2 Mt for steel and almost 48 Mt for cement [8]. Total coal supply has grown at 5.9% annually during 2000-2014 wherein coal imports have grown faster than domestic coal supply (Figure 1 and Figure 2). This has not only affected the country's trade deficit and energy security, but has also resulted in an increase in power generation costs and led to contractual disputes among producers and regulators on allowing pass through of increased cost of power generation to consumers.

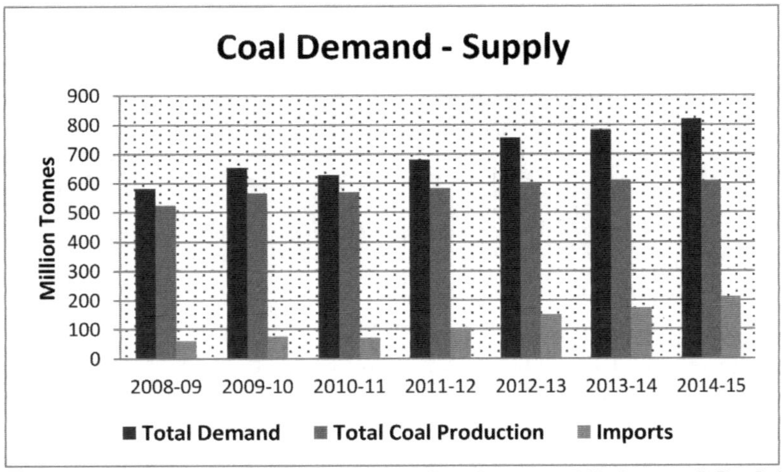

Figure 1 Coal Demand, Supply and Imports in India. Source: Ministry of Coal Annual Report (2015)

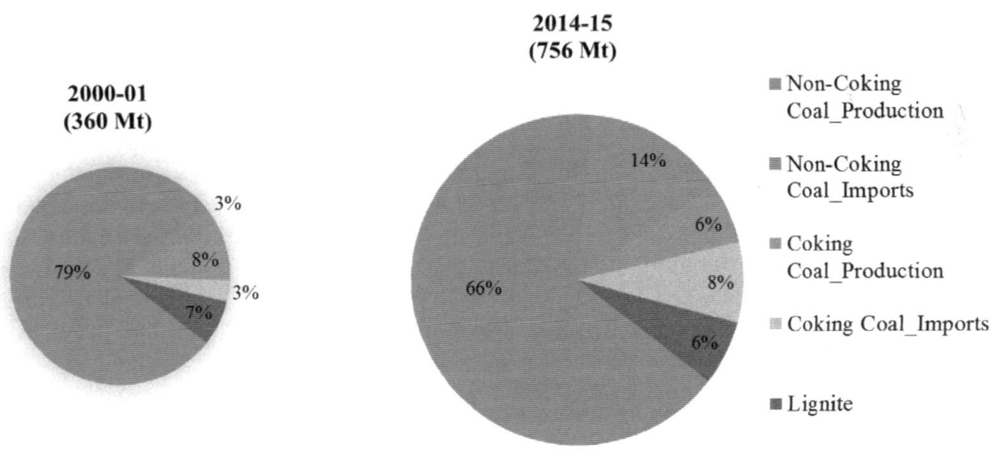

Figure 2 Coal Supply in year 2000 and 2014. Source: Ministry of Coal Annual Report (2014)

The failure to meet coal production and power generation targets over the last decade has raised many concerns for India's economic growth that need to be resolved. However, with the growing use of coal, increasing CO_2 emission is also a cause for concern. The future of coal sector would also depend on synchronizing coal use with GHG emission mitigation. This would require a detailed study of technology choices using coal and their likely emission trajectories, in order to analyse mitigation options in present and future scenarios. In this paper, a bottom-up view of the energy system using the AIM/Enduse model has been taken which is suitable for a disaggregated analysis of economic sectors for analysing energy and technology changes in different policy regimes.

3. Setting-up of the AIM/Enduse Model for the Indian Coal Sector

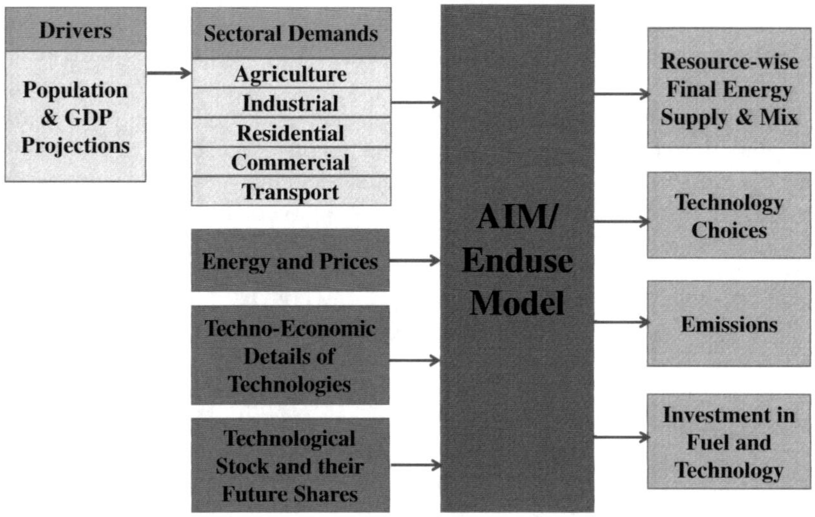

Figure 3 Model structure of AIM/Enduse model. Source: AIM/Enduse Manual (2003)

The AIM/Enduse model, a bottom-up energy and technology optimisation model, has been used in this study (Figure 3). It accounts for final energy consumption and CO_2 emissions in end-use sectors based on actual energy use and the way energy services are performed by energy devices [9]. It focuses on the end-use technology selection in energy consumption as well as energy production. It calculates the future demand of energy services for several sectors, determines the optimal set of technologies that can be used to satisfy the service demand through total cost optimisation and then it estimates future energy consumption [10]. The model minimizes the net present value of all system costs including capital, fuel, O&M and all other cost components. In this study, an analysis of the timing of fuel and technology switchover, the level of emission and different ways of mitigating it by adopting countermeasures has also been carried out. This has helped in assessing the requirements for transition in the coal sector in view of achieving mitigation benefits while providing continuous energy security.

The model facilitates the analysis of the energy consumption, technology choices, emission trajectories etc. in a given context using linear optimisation. [11, 12]. We have set up the country-level AIM/Enduse model for India that requires selection of sectors, services, technologies, reference years and discount rates. In this study, we have taken up five commonly classified sectors, namely, the industrial, transportation, agriculture, residential and power sectors for the bottom-up modelling analysis [3, 4, 13]. Also one additional sector namely the 'coal sector' has been added for specialised treatment of coal and provided a detailed description of the coal sector, since coal is a primary energy source in various intermediary and end-use technologies. Multiple services in each sector have been taken for better understanding of the sector. Most technological competition can be described as a substitution of one technology for another. The technologies considered in the model range from the ones currently available to those that are still in the research and development stage.

Service demands are the main external drivers that trigger technology selection decisions based on information on costs. Energy-mix and material-mix are derived from the technology-mix. Finally, CO_2 emissions are derived from the information on the emission characteristics of energy, materials and technologies. The service demand estimates have been done exogenously. The methodology for demand projection used by Kapshe et al [4] has been adopted in this study.

The maximum share in the model is the estimation of the future scenario in the sector. The maximum shares for the base year are taken from various government and research publications. The model provides enough scope for accommodating changes in figures on demand, shares, etc. as demand elasticity changes or a particular policy push is observed in due course. The coal based power generation technologies included in the model are of four generic types – Sub critical PC, Super-critical PC, IGCC and Fluidized Bed Combustion. Their fuel efficiency and CO_2 output assumptions are indicated in Table 1.

Table 1. Performance of coal-based technologies

Parameter	Year	Technology			
		Sub-Critical	Super-Critical	IGCC	Fluidized Bed
Coal (kg/kWh)	2005	0.616	0.568	0.549	0.549
	2030	0.494	0.474	0.457	0.457
	2050	0.441	0.431	0.416	0.416
Emissions (kg-CO_2/ kWh)	2005	1.159	1.069	1.032	1.032
	2030	0.930	0.891	0.860	0.860
	2050	0.829	0.810	0.782	0.782

3.1. Scenario narratives

The continuing trend of development in India is energy intensive. Currently, the carbon emission limits have not been implemented in India, however, there is a high likelihood of its imposition based on the envisaged high growth in future. The model development has been done in commensurate with on-going measures and future targets. Alternate strategies and development path could be studied through various scenarios. The Business-as-Usual (BAU) scenario in the current study includes INDC's and reflects the continuance of policies and practices with moderate changes. It is an attempt to consider the role of advanced technologies in achieving the targets of energy security and climate security together. This goal in continuance with on-going efforts will require impetus on necessary sector specific actions to reduce emissions intensities.

The policies have been selected on the basis of their potential to reduce carbon dioxide emissions. The present study considers only the BAU scenario for analysing the fuel, technology and emission trends. However, other scenarios may also be worked upon which will further bring down the emissions. Our modelling also captures the latest commitments by the Government of India in its Intended Nationally Determined Contributions (INDC) report to the UNFCCC on October 1, 2015 [14]. The INDC initiatives on GHG mitigation and clean energy cover the period 2005-2030.

Clean coal constitutes a major part of these initiatives. INDC (2015) [15] indicates that "coal based power as of now accounts for about 60.8% (167.2 GW) of India's installed capacity. In order to secure reliable, adequate and affordable supply of electricity, coal will continue to dominate power generation in future. Government of India has already taken several initiatives to improve the efficiency of coal based power plants and to reduce its carbon footprint. All new, large coal-based generating stations have been mandated to use the highly efficient supercritical technology. Renovation and Modernization (R&M) and Life Extension (LE) of existing old power stations are being undertaken in a phased manner. About 144 old thermal stations have been assigned mandatory targets for improving energy efficiency. Coal beneficiation has been made mandatory. Introduction of ultra-supercritical technology, as and when commercially available, is part of future policy. Besides, stringent emission standards being contemplated for thermal plants would significantly reduce emissions." INDC has also identified Pulverized Combustion Ultra Super Critical, Pressurized Circulating Fluidized Bed Combustion, Super Critical, Combine Cycle, Integrated Gasification Combined Cycle and Fuel Cell as priority clean coal technologies for India. These technologies are well represented in our model.

Our results provide an analysis of Indian coal futures under alternate scenarios during 2000-2050 and therefore we divide our analysis for the 2005-2030 period to synchronise with Indian INDCs. The other future period in our analysis for 2030-2050 provides the long-term coal scenario for India.

4. Results and discussion

An analytical interpretation of the quantitative outputs of the model for the BAU scenario has been presented in this section. The BAU scenario has been generated by running the model from the year 2000 to 2050. The results obtained in the BAU scenario indicate that huge capacity expansion is needed across sectors fuelling energy demand at a high rate. This shows that most power generation utilities continue to use coal and expand their coal-based generation capacity to meet the increasing energy demand. Given the emphasis on energy security concerns and the requirements of meeting the energy demand, the use of coal is inevitable in estimated visions. New generation options like Atmospheric Fluidised Bed Combustion (AFBC), Pressurised Fluidised Bed Combustion (PFBC) and Integrated Gasification Combined Cycle (IGCC), etc. are advocated because of their higher generation efficiency which imposes a lower environmental burden and their ability to work with any type of coal. Super-critical technologies are also preferred due to their higher efficiency compared to conventional sub-critical pulverized coal technology [16-19].

4.1. Fuel-technology mix

Indian energy needs revolve around coal, ranging from power generation to industrial production, and for cooking purposes in rural areas. The total energy consumption in India increases by about 3 times over 2005-2050 (Figure 4). Because of reduced availability and increasing access to commercial fuels, the biomass consumption will decrease, however the use of commercial biomass and biomass is observed to increase in power and transport sector. Diesel and gasoline have also registered a substantial increase due to the rapid growth of the transport sector and cumulatively account for nearly 22% by 2050. The share of gas increases from 7% to 18% in 2050, thereby substituting coal due to increasing availability through international collaborations mainly in power, industry and transport sectors. The share of coal is expected to remain almost constant due to increasing reliance on coal based electricity generation and domestic availability. However, the share of coal is observed to decrease due to increase in share of gas, nuclear and renewables in the future energy mix.

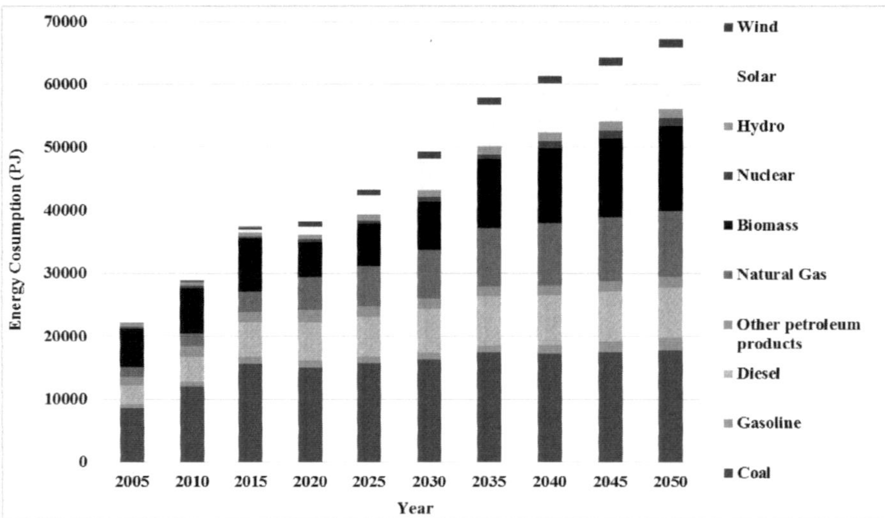

Figure 4. Fuel-wise primary energy consumption in end-use sectors in BAU

Electricity production has been the major source of carbon dioxide emissions in India. Most of its current and future requirement is still expected to be based on coal and gas in addition to aggressive targets for renewable energy. This is mainly because given the scale at which the demand is growing, no single scheme and promotion policy can fulfil the demand generated across sectors [20]. The technology-wise electricity generation has been depicted in 2050 in India (Figure 6). Total electricity generation increases at an almost constant growth rate of 2.3% annually from 2015 to 2050. The share of coal-based technologies in electricity generation is constantly declining from around 70% in 2005 to 43% in 2030 due to inter-fuel substitution and higher availability from renewable and further decreases to 28% in 2050. The corresponding increase has been observed in gas-based generation from around 10% to 21% in 2030 and 25% in 2050. But coal still remains the mainstay for power generation followed by solar, gas, wind and hydro.

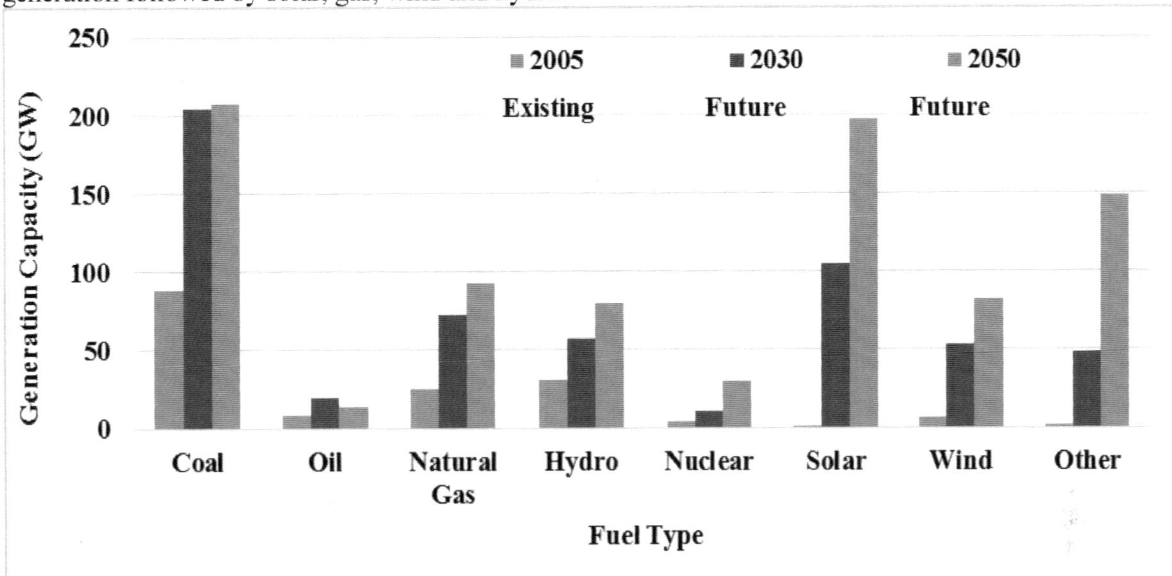

Figure 5. Existing and future generation stock in BAU

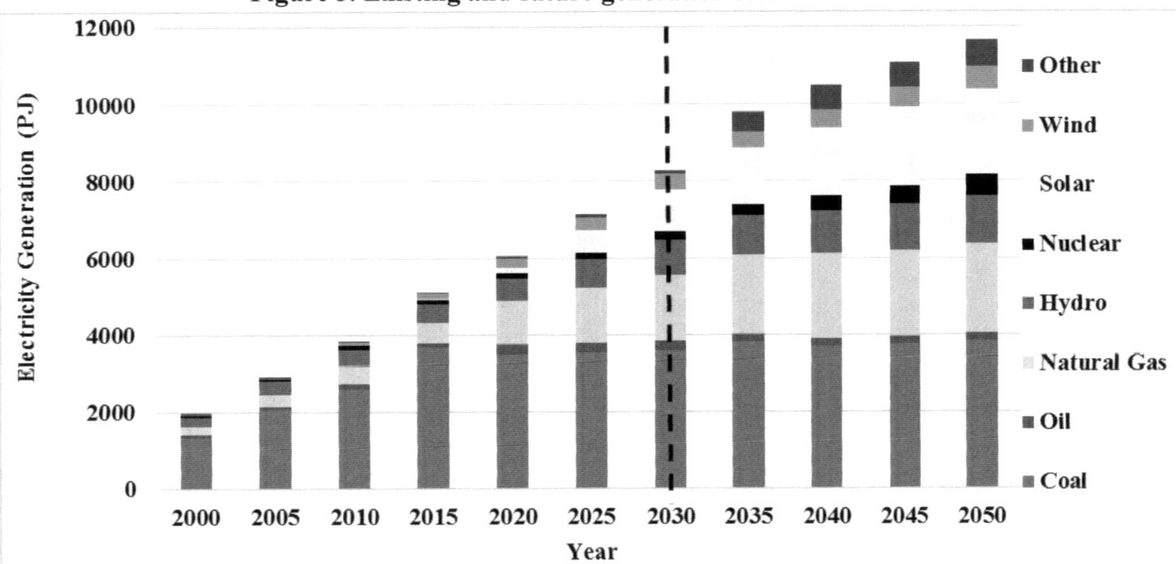

Figure 6. Technology-wise electricity generation in BAU

The total installed capacity goes up nearly five times, 568 GW in 2030 in the BAU case. The share of coal based capacity is declining from 53% to 35%, while gas based capacity marginally increases from 9.9 % to nearly 12.9% over the span of 30 years (Figure 5). There is an increase in the coal capacity till 2020, after which the old, inefficient thermal power plants will be phased out. These power plants will be replaced by efficient power plants that will use a combination of improved pulverization along with super-critical and ultra-super critical technologies. Coal beneficiation and fly ash utilization will be required to mitigate other environmental hazards of enhanced coal use in this scenario (Figure 5). Gas based generation would compete with coal in future among fossil fuel based technologies. Creating new large hydro capacity will become a challenge due to strict land acquisition laws, and relocation and resettlement issues. Climate change is creating further stress on the overall water availability in the peninsular rivers. As a result the hydro capacity share continues to fall from around 22% in 2000 to 10% in 2030, and much below beyond that. GoI has announced the establishment of 100 GW of solar capacity by 2022, and augmentation of total wind capacity to 60 GW by 2022 [15].

The generation capacity of renewables (solar, wind, small hydro and biomass) accounts to around 40 per cent in 2030 and around 50 per cent in 2050. In the case of increasing concerns for limiting thermal generation, nuclear technologies emerge as a promising choice for meeting energy needs. However, the 'Fukushima' nuclear disaster in Japan has brought the discussion on nuclear safety to the forefront and there is a developing resistance in various countries to nuclear power. Given the current concerns and slow progress, nuclear power is expected to contribute only around 2-3% of total generation in the BAU scenario, thus limiting its scope in the BAU scenario. The present energy generation stock and plausible capacity addition in year 2030 and 2050 respectively is shown in Figure 5. It indicates that coal technologies have a major share in power generation followed by gas among fossil fuel-based generation. More than 24% of new installed capacity comes from coal while around 11% and 9% respectively is added by gas and hydro. Solar and wind generation capacity is envisaged to be more pronounced due to the policy thrust and likely to contribute around 24% and 10% over the period of 45 years to meet future requirements.

If we carefully look at Figure 5 and Figure 6, we observe that coal based capacity has grown at a CAGR of 2.1% annually over the span of 45 years whereas electricity generation from coal based technologies has increased at a CAGR of 3% during the same period. Also the share of coal technologies in generation has declined from 59% to 22.8% from year 2000 to 2050 while coal consumption in generation for the same period has declined from 69.5% to 27.5%. This is primarily due to the energy efficiency improvement of coal technologies and emerging competing technologies. The Union government has launched the Restructured-Accelerated Power Development and Reforms Programme (R-APDRP) to improve efficiencies across the generation sector and minimize wastage. The T&D losses have significantly improved from about 32% in 2005 to 23% in 2013. This has improved the resultant power supply by reducing the wastages. Roughly India's coal requirement is 0.7 kg/kWh. So, avoiding 10 per cent of electricity being generated saves around 70 million tonnes of coal consumption per year [21]. Natural gas consumption in BAU has increased from 10.8 % to 25% over fifty years. This inter-switching between coal and gas seems possible in view of improving international trade practices and technology improvements in transporting networks for natural gas in the future. Gas-based capacities have grown in the past and are preferred by captive producers due to their low-cost and ease of operation [22]. Bilateral contracts and agreements among countries are providing a more robust business foundation and avoiding any contingencies and disputes to a great extent. It is expected that this would facilitate strong tie-ups among countries for signing future long-term contracts.

4.2. Coal-based power generation mix

The new coal technologies improve fuel efficiency and mitigate carbon emissions at three different stages, namely pulverization of coal, production of steam and sequestration of carbon dioxide. Coal is pulverized by a combination of crushing and grinding technologies to get a desired degree of fineness of the fuel, which in turn increases the fuel efficiency by 70-80 per cent. The sieved coal is fed into a furnace to produce steam. The technologies used to pulverize coal include cyclone-fired boilers, various types of fluidized bed combustion and integrated coal gasification/combined cycle technologies. The efficiency of heating unit is increased by improved thermodynamics when water is converted to superheated steam without boiling. Super critical and ultra-super critical technologies produce steam and higher temperature and pressure, which enable more efficient operation of the turbine cycle as compared to its sub-critical counterparts. The mitigation of carbon emissions at the end of the cycle can be achieved by addition of sequestration technologies. A number of technologies exist that capture and store carbon dioxide from gas stream, however they have not yet been optimized for the scale required to sequester the carbon in India. Apart from the conventional technologies, a combination of aforementioned pulverization, heating and sequestration technologies have also been considered in the model. The new technologies and their introduction year are based on various Indian national and international reports and expert opinions from the government and industry.

The details of coal based generation technologies and their expected capacity addition in different timeframes has been given in Table 2. It clearly indicates that the share of coal based technologies have to increase in BAU scenario since coal based power plants, with their relatively higher plant load factors (PLF) vis-à-vis renewable technologies, fulfil the maximum power demand. Fluidized bed combustion (PFBC, AFBC, CFBC) and IGCC also increase their respective capacities in future. India has already established 27.5 GW of supercritical units and around 50 GW such capacities are under construction [15].

Table 2 Coal-based technology-wise generation capacity (GW) in BAU

Year	Sub-critical	IGCC	Fluidized Bed	Super-critical	Lignite	Coal based total capacity (GW)	Coal as % of total national generation capacity
2005	79.9	3	2.1	0	3.7	88.7	53.8
2015	128	4.9	5.3	12.1	3.8	154.1	55.9
2020	138.5	5.6	8.6	55.3	4.8	212.8	47.9
2030	111.1	8.7	12	81.9	5.7	219.4	38.4
2040	89.3	11.8	12.7	109.8	5.7	229.3	30.2
2050	67	15	22	123.9	4.2	232.1	27.2

4.3. Coal demand

The coal demand in India would continue to rise at a 1.62% CAGR reaching about 1066 Mt in 2050 *vis-a-vis* 516 Mt in 2005 (Figure 4). But due to penetration of advanced technologies in coal intensive sectors and fuel switching, this demand may likely to go down. This analysis of alternate possibilities will be the subject matter of next part of the study.

However, considering the vast geological reserves possessed by India and the government focus on low cost energy security, it is clear that coal will continue to be the mainstay energy choice for the coming few decades. Simultaneously, there is pressure from the social and government sectors to use coal in a cleaner and more efficient way. This can be achieved in one of three ways - providing clean and washed coal to industries (pre-use), using higher efficiency technologies (in-use) and capturing emissions (post-use). On the other hand, the Power and Coal Minister has indicated his plans to set up coal washeries for 500 Mt of coal every year for the next five years, indicating a total coal handling of over 2.5 billion tonnes per year by 2020 [23].

However, this raises the concern of creating massive infrastructure and relocating inhabitants, as many infrastructure projects have been contested in India lately. Project execution deficiencies reflected in slower government approval processes and resistance by social activists have only added to turning down or closure of projects. Therefore, in this tight supply situation, achieving indigenous coal production capacity is daunting. Many of the new thermal coal based capacities propose to use imported coal (mandatory bidding conditions for the power sector in some UMPP's) have led to the creation of increasing port capacities to facilitate coal imports. In view of these targets, measures to augment coal supply need greater thrust.

4.4. CO_2 emissions

The CO_2 emissions from end-use sectors in India have increased from around 1545 Mt-CO_2 in 2005 to nearly 4183 Mt-CO_2 in 2050.

Despite efficient vehicular stocks and marginal fuel substitution, the transport sector emissions have increased. In the residential sector, fuel switching from kerosene to LPG and electricity has had a considerable positive impact on carbon emissions. However, a majority of emissions around 44.5 % and 28.1 % are from the industrial sectors and power generation respectively in 2050. Coal is a major energy input in both the sectors (Figure 7).

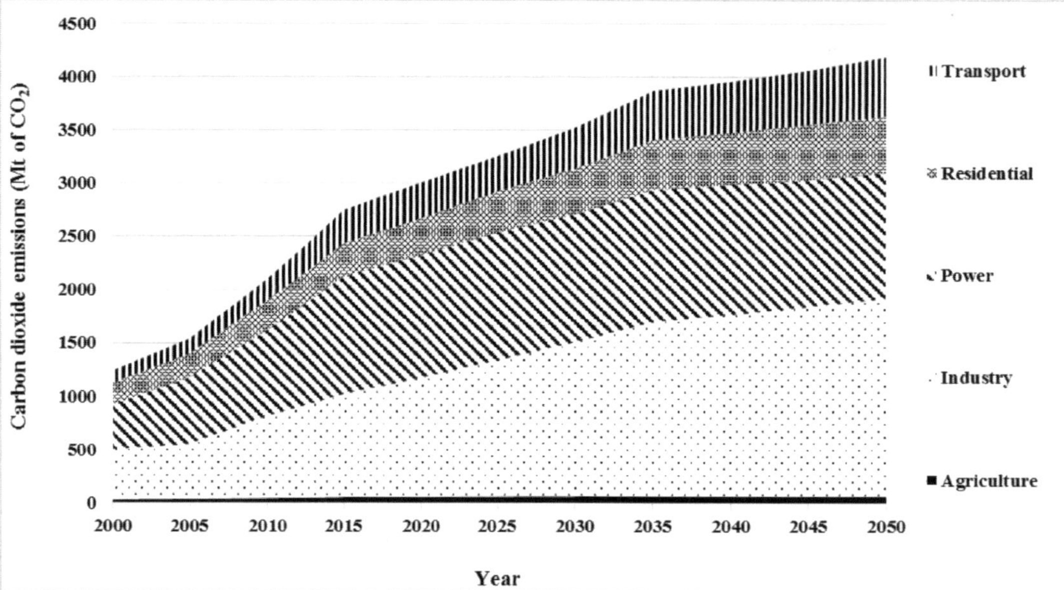

Figure 7 CO_2 emissions from end-use sectors in BAU

Total coal demand has increased from around 9117 PJ in year 2005 by twice in 2050. This shows that coal will serve the major needs of energy in India and will remain the main contributor to rising CO_2 emissions. It is apparent that any efforts at reducing energy related emissions will have to consider the effect of increasing coal use.

5. Conclusion

In order to see the organised responses for tackling climate challenges in future, climate friendly policies are necessary. These policies would be enacted in due course to realize mitigation benefits. Energy security and climate change mitigation have become important in the changing global landscape of energy policy formulation and lead to identifying solutions that can survive in the long run. In the BAU scenario the requirement of coal is continuous and coal based capacity contributes nearly half of the total generation. Its share declines by almost 12% in 2030 and further by 15% in 2050 as compared to its share of 63% in 2005 though coal capacities continue to increase at a CAGR of 2.78% till 2050. Due to the continuing thrust on climate change mitigation as committed in the Indian INDC, the role of clean coal technologies would be decisive in future along with other carbon neutral technologies such as nuclear and renewable that are expected to contribute 7.5% and 13.3% respectively to total generation by 2050.

The overall change in energy mix (fossils, nuclear and renewable) under various alternate possibilities would depend upon the intended target actions decided by the Government for future growth and any GHG mitigation targets that may arise due to stricter global climate change regimes.

A major thrust is required to create a favourable policy climate, technology R&D and transfer, and international financial support for large scale penetration of clean coal technologies in India to ensure energy security with GHG mitigation.

IOP Conf. Series: Earth and Environmental Science **76** (2017) 012001 doi:10.1088/1755-1315/76/1/012001

References

[1] MoSPI 2015 *ENERGY STATISTICS 2015* (New Delhi: Ministry of Statistics and Programme Implementation, Central Statistics Office, Government of India) Retrieved from http://mospi.nic.in/Mospi_New/upload/Energy_stats_2015_26mar15.pdf. 2015

[2] Garg A and Shukla P R 2009 *Energy* **34** 1032–1041

[3] Garg A, Kapshe M, Shukla P R and Ghosh D 2002 *Atmospheric Environment* **36** 213–224

[4] Kapshe M, Garg A and Shukla P R 2003 *Climate Policy Assessment*, ed M Kainuma, Y Matsuoka, T Morita (Tokyo: Springer Japan) p 93-112

[5] Nair R, Shukla P R, Kapshe M, Garg A and Rana A 2003 *Mitigation and Adaptation Strategies for Global Change* **8** 53-69

[6] Garg A, Shukla P R and Kapshe M 2006 *Atmospheric Environment* **40** 4608–4620

[7] DOE 2015 U.S. Department of Energy. CDIAC. Carbon Dioxide Information Analysis Centre. (2. G. Project, Producer) doi:DOI: 10.3334/CDIAC/GCP_2013_V2.3

[8] MOC 2011 *REPORT OF THE WORKING GROUP ON COAL AND LIGNITE* (New Delhi: Government of India

[9] Morita T, Kainuma M, Harasawa H and Kai K 1996 A Guide to TheAIM/ENDUSE Model - Technology Selection Program with Linear Programming (Japan: AIM Interim Paper)

[10] Selvakkumaran S and Limmeechokchai B 2015 *Energy Policy* **81** 199-214

[11] Hibino G, Pandey R, Matsuoka Y and Kainuma M 2003 *Climate Policy Assessment*, ed M Kainuma, Y Matsuoka, T Morita (Tokyo: Springer Japan) p 247-283

[12] Fujimori S, Kainuma M, Masui T, Hasegawa T and Dai H 2014 *Energy Policy* **75** 379–391.

[13] Mallah S and Bansal N K 2010 *Energy Policy* **38** 1059–1066

[14] GOI 2015 INDIA'S INTENDED NATIONALLY DETERMINED CONTRIBUTION. UNFCCC. Retrieved from http://www4.unfccc.int/submissions/INDC/Published%20Documents/India/1/INDIA%20INDC%20TO%20UNFCCC.pdf

[15] INDC. Retrieved from TIMES OF INDIA: http://timesofindia.indiatimes.com/home/environment/developmental-issues/Indias-INDC-ignites-hope-for-a-green-future/articleshow/49226076.cms. October 5, 2015.

[16] Gopinath V K, Balagurunathan S and Gopalakrishnan V 2002 *Advanced clean coal technology, perspectives for India*. See also, <www.igeos.pl/doc/coaldialogue/ WECCoalSlides.ppts>

[17] Brown M A, Levine M D, Short W and Koomey J G 2001 *Energy Policy* **29** 1179–1196

[18] TERI 2006 *National energy map for India: technology vision 2030* (New Delhi: The Energy and Resource Institute)

[19] Mallah S and Bansal N K 2011 *Applied Energy* **88** 622-629

[20] Thiam D R, Benders M J R and Moll C H 2012 *Applied Energy* **94** 98-108

[21] CEA 2015 *September. Executive Summary* Retrieved from http://www.cea.nic.in/reports/monthly/executivesummary/2015/exe_summary-09.pdf

[22] Bhattacharya M, Rafiq S and Bhattacharya S 2015 *Applied Energy* **154** 686–695

[23] TOI 2015 September 13, 2015. Retrieved from Times of India: <http://timesofindia.indiatimes.com/home/environment/developmental-issues/LED-bulbs-to-save-up-to-Rs-40000cr-says-Piyush-Goyal/articleshow/48941146.cms>

Acknowledgement

The authors acknowledge the financial support received from "Comprehensive, Whole-of-Government Strategies on Coal Transition in Major Coal Using Countries" project by KR Foundation and Sub-Grantee IDDRI that supported part of this research.

IOP Conf. Series: Earth and Environmental Science **76** (2017) 012002 doi:10.1088/1755-1315/76/1/012002

Geological and Rock Mechanics Perspectives for Underground Coal Gasification in India

Ajay K Singh*, **Rajendra Singh**

Strata Mechanics and Nonconventional Gases Group, CSIR-Central Institute of Mining and Fuel Research, Dhanbad, India

* Corresponding Author. Email: ajay.cimfr@gmail.com

Abstract. The geological resources of coal in India are more than 308 billion tonnes upto a depth of 1200 m, out of which proved reserve has been reported at around 130 billion tonnes. There is an increasing requirement to increase the energy extraction efficiency from coal as the developmental prospects of India increase. Underground coal gasification (UCG) is a potential mechanism which may be utilized for extraction of deep-seated coal reserves. Some previous studies suggest that lignites from Gujarat and Rajasthan, along with tertiary coals from northeastern India can be useful from the point of view of UCG. We discuss some geological literature available for these areas. Coming to the rock mechanics perspectives, during UCG the rock temperature is considerable high. At this temperature, most empirical models of rock mechanics may not be applied. In this situation, the challenges for numerical modelling of UCG sites increases manifold. We discuss some of the important modelling geomechanical issues related to UCG in India.

1. Introduction

Stringent climate targets are likely to change the way of coal usage in the near and long term. With an view to strengthen India's energy security, an effort must be made to diversify coal usage while also finding potential mechanisms to reduce our natural gas imports. This becomes even more relevant for India as 55% of the commercial energy comes from coal and lignite. This trend is likely to continue for the years to come as per various modelling studies and expert elicitations [1, 2]. In this state of affairs, coal gasification may become a useful technology for India.

Surface and underground coal gasification both are at the forefront of clean coal technologies. It is demonstrated that coal reacts with some oxidants like oxygen or air and sometimes steam at elevated temperature and pressure in a reactor to release a product synthetic gas (syngas), which is a mixture of carbon monoxide, hydrogen, methane, carbon dioxide, nitrogen, hydrogen sulfide and other trace gases. The product syngas when used in an Integrated Gasification Combined Cycle (IGCC) turbine, provides significantly high efficiencies for power generation. Capture and storage of carbon dioxide, which is another topic of major research interest throughout the globe, is amenable to gasification of coal. The essential concept of underground coal gasification (UCG) also known as in-situ gasification is to develop an underground gasifier. Like surface gasifiers, UCG process also involves simultaneously existence of oxidation, reduction and distillation zones in the underground gasifier.

Majority of the UCG trials throughout the globe has been attempted using the linked vertical wells method in flat or gently dipping coal seams [3]. In this method, a pair of boreholes called injection and production wells are drilled from the surface into the horizontal coal seams. Linking is established between the two boreholes through the coal seam in order to ease the flow of gases from the injection well to the gasification zone and also from the combustion zone to the production well. The coal is ignited at the bottom of the injection well. Gasifying agents such as air or enriched air or oxygen and steam are injected into the coal seam through the injection well to support ignition followed by gasification of coal.

Energy demand of India is increasing continuously and being the major fossil fuel, coal plays an important role in the energy sector. India is the third largest producer of the coal in the world. The geological resources of coal in India is reported as more than 300 billion tonnes upto a depth of 1200m as on 01.04.2015 out of which proved reserve has been reported as 125.90 billion tonne [4]. The coals are of sub- bituminous to bituminous rank and in some places lignite, having high ash content and low calorific value, which makes it uneconomical to mine from such a greater depth. In India, UCG was first taken up in the mid 1980's. The Government of India has already indicated production of syngas obtained through UCG as "Approved End Use". Hence, in addition to the implementation of a pilot UCG projects, two coal blocks namely Kaitha in Ramgarh coalfield and Thesgora 'C' in Pench-Kanhan valley coalfields were identified for commercial development.

This paper discusses some of the geological and rock mechanics perspectives related to UCG projects in India

2. Rock Mechanics Perspectives

UCG processes are likely to lead in deformation of 60% strata (if we assume a coal with 40% ash content). In conventional mining, all the material would be extracted, but in UCG this reduces by a margin of 40%. By removal of excess material in conventional mining processes, there is considerable deformation in the strata (Figure 1). This shows an advantage in terms of the residual ash and also in the rock mechanics perspectives. However, in case proper geologic controls are not in place, there is a fear of subsidence. Thus, it is imperative to predict the amount of displacement faced by the rock strata [5]. Modeling can easily indicate the amount of physical displacement of the strata but it may not be useful in defining the geo-engineering characteristics of the rocks. Since, high temperature causes drastic variations in the rock properties; the modeling of UCG is considerably difficult.

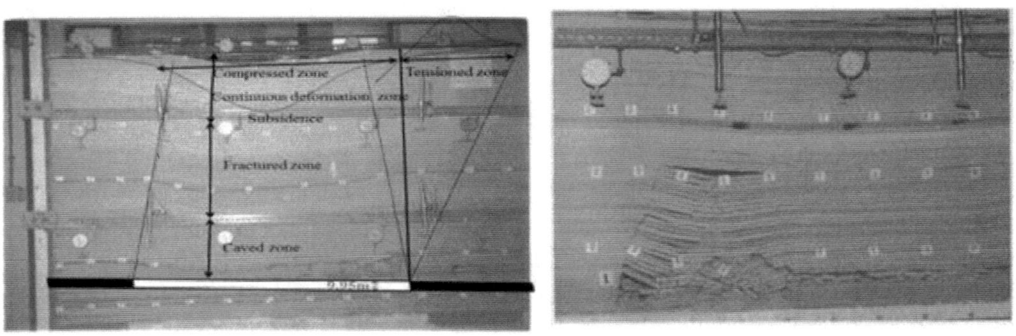

Figure 1. Simulation of deformation and displacement of rock strata using rock-mechanics based computational techniques (after [6])

Both physical and numerical models can be utilized to understand the strata mechanics patterns during excavation of caverns [6, 7] . In such studies, it has been found that the depth as well as the width of the tunnel/cavern has significant impacts on the strata mechanics (Figure 2), in addition to the overlying rock strata properties (Figure 3). Mathematical modelling in such cases is a convenient way to model rock mechanics during mining operations. The results of such models can be then compared to empirical results to ensure that the developed models suit the purpose of our studies, and can be constantly improved taking field considerations into account. Thus, we zero-in on the suitable models for simulating different cases of mining operations. A number of such empirical rock failure criteria have been developed, an overview of which may be found in Sheorey [8]. Researchers have suggested that the results found in empirical studies, may be significantly different from established international models. For instance, Sheorey et al [9] have pointed out that there are considerable variations in the field studies carried out for Indian coalfields, as compared to the Subsidence Engineers Handbook, UK. They have also demonstrated that difference exist in the cases of extraction below unfractured or uncaved rock and below pre-existing caved goaf(s). Some studies have been developed for estimating subsidence behaviour in UCG [10, 11]. Geophysical monitoring studies have also been suggested for UCG [12].

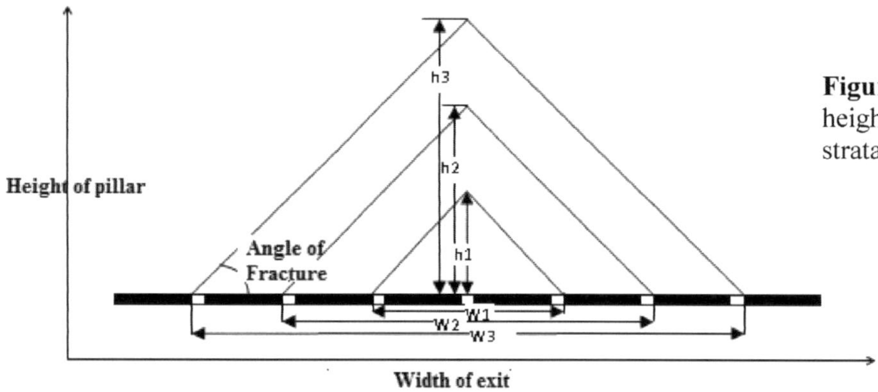

Figure 2. Effect of height and width on strata mechanics

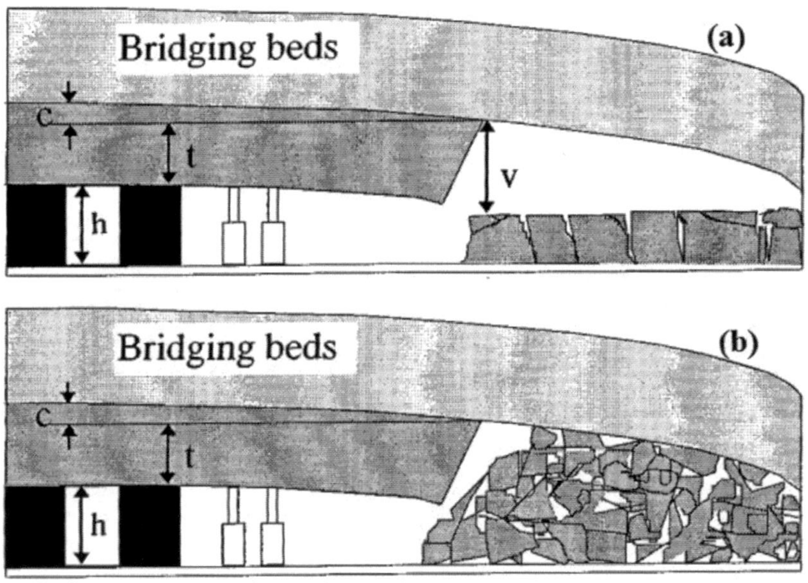

Figure 3. Effect of rock properties for the case of (a) strong and (b) weak overlying rock strata ([6])

In fact, presence of massive rock and complex geo-mining conditions induce large complexities into modelling rock mechanics problems. This raises the imperative for some simplifications, so far as possible. In a physical model, the process of assimilating the equivalence of strength by constantly incorporating the condition and equality of the event is inherently contained. The numerical method shows the methods that are direct attempts to solve fundamental differential equations, integral equations, or the principle of change. With advances in computational power, mathematical models for rock mechanics problems are becoming significantly popular, and a number of recent works have come up in this area [13]. Analysis of results obtained from efficient modelling can greatly reduce the costs and provides complete information about the distortion of rock strata. Most of the information in geotechnical models is extracted from the coal core obtained from the boreholes. These cores are tested for mechanical strength in the laboratory. The results from small core samples are translated into the results for the entire rock mass. However, in many cases, these results do not match, especially in case of massive rocks. To avoid this uncertainty, modelling results are compared with the results of empirical formulation.

As stated, for the modeling of UCG, defining physical and engineering properties of large rocks becomes quite complicated as the high temperature has significant impact on the physical and engineering properties of rocks. To our knowledge, no integrated study so far exists, which estimates the changes in rock properties in such high temperatures for Indian coalfields. Luo and Wang [14] have estimated the impacts of high temperature on the properties of mudstone in the UCG process. Yu et al [15] have also studied the development of micro-pores in coal structure with temperature. Liu and Xu [16] have estimated the threshold temperatures for change in strength for granite using longitudinal wave velocity. Similar studies should be carried out for Indian coalfields for assessing feasibility of UCG in India. In fact, Sirdesai et al [17] have carried out a computational study using COMSOL Multiphysics software to understand subsidence behaviour during UCG. They have concluded that cavities of width less than 100 m cause lower subsidence. This study can further be improved with addition of geological constraints such as presence of groundwater and other features. Figure 4 shows the effect of depth and cavity width on the degree of subsidence.

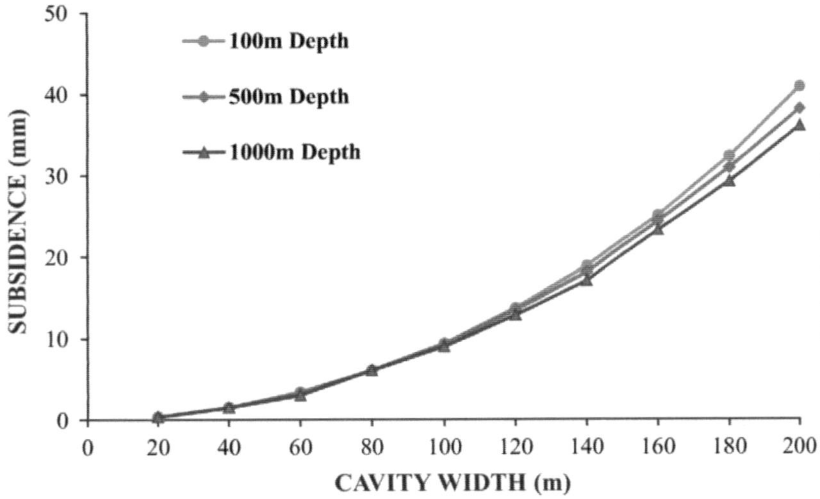

Figure 4. Effect of cavity width and depth on subsidence. Source: Sirdesai et al [17]. Reproduced with permission from authors

In general, there is no need to account for temperature variation in conventional mining processes. However, UCG is an exception to this as the rock temperature may reach in excess of 1200 ^{0}C. This temperature can be maintained after a long period of time after combustion of coal and continuously during gasification. The deformation process of rock at high temperatures depends on contact with mineral structure and deposited grain [18]. At high temperatures, properties are affected majorly due to the deposited particles in the rocks such as soil, clay, sand etc. Quartz-Feldspars minerals present in the rocks affect the rock properties [19]. Malkowski [18] has also demonstrated that various rock respond differently to high temperature based on their depositional features.

3. Geological perspectives

In the past, our group has focussed on UCG in Cambay Basin (Gujarat), Barmer-Sanchor (Rajasthan) and north-eastern coalfields. Chemical analysis of lignites from Gujarat show that fixed carbon content is between 30-45%. Further, it was found through thermo-gravimetric analysis (TGA) that Umarsar lignite is the most reactive in this region, followed by lignites from Vastan, Panandhro, Valia and Bhavnagar [20]. Syngas modelling was also carried out for lignites from Kapurdi mine in Rajasthan. These are coals with high moisture and low sulphur content. The study suggests that at 20% oxygen concentration, some amount of coal is left unreacted while at 60% concentration, the CO_2 formation is

IOP Conf. Series: Earth and Environmental Science **76** (2017) 012002 doi:10.1088/1755-1315/76/1/012002

higher than CO [21]. We also performed an integrated project relating to the feasibility of UCG in the northeastern India, the full report of which may be found in [22].

Some other studies with relevance to UCG may be discussed here. Daggupatti et al [23] have shown using modelling that in lignites, that bypassing of oxygen may occur in cases where the cavity in the forward direction closely approaches the production well. Prabhakar et al [24] have performed modelling studies for sub-bituminous coal and concluded that combined kinetics and heat transfer model is more effective in prediction gasification characteristics. Singh et al [25] have predicted that for heat affected coals from the Jharia coalfield, changes in properties are similar for in-situ or laboratory carbonation. Mukherjee and Srivastava [26] concluded that when heated in presence of air, the stretching, bending bands disappear. A highlight of some other studies focussing on geology and petrography in Cambay Basin, Barmer-Sanchor and Northeastern coalfields is shown in Table 1.

Reference	**Methodology/Assumptions**	**Results**
Gujarat		
Singh et al [27] (*Petrological and chemical studies on Vastan lignites in Gujarat*)	• Samples from open mines were collected using pillar sampling • Leitz Orthoplan Pol microscope with an Photoautomat MPS 45 control module was used for microscopic study • Particle size for microscopic analysis was kept at 18 mesh size • 70 mesh size particle size was used for proximate and ultimate analysis	• Out of huminites, liptinite and inertinite maceral groups, huminites are in abundance • Fuginities occur individually and in clusters • Lignites are reported with high moisture, moderately high volatile matter and varying ash content. • Average sulphur content is 1.59% • Sulphur content is more for lower seams • Sulphur content increases with increase in ash content
Singh et al [28] (*Petrological and chemical studies on Rajpardi lignites in Bharuch district in Gujarat*)	• The samples were collected from Amod mine in Jhadagia area in Bharuch • Pillar method of sampling as used • Microscope with an Photoautomat MPS 45 control module was used for microscopic study • Particle size for microscopic analysis was kept at 18 mesh size • 70 mesh size particle size was used for proximate and ultimate analysis	• Huminites were in abundance ranging from 68.9 to 92.6% in vol basis • Telohuminite and detrohuminite are a majority within humnities • Liptinites ranged from 1% to 6.5% in vol. basis • Mineral matter content showed large variation • Lignite was identified with have high volatile matter and hydrogen content
Rao et al [29] *(Palynostratigraphy and study of depositional environment of Vastan mine, Gujarat)*	• 33 samples were collected of which 31 were finally used • Palynofossils were recovered from these samples • Samples were treated with	• 86 generea and 105 species of alagal remains were identified • 7 generea and 8 species of fungal remains were

	HCl, HNO_3 and HF and later with 5% KOH	indentified • 57 generea and 73 species of angisperms remains were identified • Three cenozones zones namely Proxapertites spp. Cenozone, Operculodinium centrocarpum Cenozone and Spinizonocolpites spp. Cenozone were identified on the basis of percentage frequency of palynomorphs
Mallick et al [30] (*Pyrolytic and Spectroscopic studies on Eocene resin in Vastan lignite*)	• FTIR spectroscopic analyses were done in transmission mode on a Nicolet MANGA 550 model running OMNIC software • The sample was flash pyrolysed at 600°C for 20 sec by a CDS analytical pyroprobe • Pyrolysis chamber was kept at a constant 300°C • Curie point pyrolysis was conducted in a Curie point pyrolyser	• Fossil resin contains 78% of C, 10% of hydrogen by weight • Fossil resins were found to be rich in aliphatic compounds from the FTIR data • Cadalene based C-15 bicyclic sesquiterpenoids were found in abundance • Upon pyrolytic study, Vastan resin are identified as class 2 resin
Rajasthan		
Singh et al [31] (*Thermal maturity and hydrocarbon potential studies on lignites of Bikaner-Nagaur basin*)	• Sampling was done using pillar coal sampling technique • Sample was prepared at 18 mesh size for petrographic analysis and 72 mesh size for proximate and ultimate analysis • Maceral analysis was conducted at BHU, using a Leitz Orthoplan Pol Microscope • Hydrocarbon potential determination was done using Rock Eval-6 pyrolysis	• Lignite was found to be rich in huminite macerals (86.3% on an avg in mmf basis) • It majorly contains kerogen organic matter • Liptinite content was 4.4% on an average • Mineral matter shows variation • Lignites have high volatile matter content (54.4% to 60% in daf basis)
Mukherjee et al [32] (*Study of physical, chemical and petrographic properties of Kapurdi lignite in Barmer Basin*)	• Study was conducted on Kapurdi lignites deposited in Barmer Basin, Rajasthan.	• Vicker micro-hardness parameter values varied between 20 to 22 kg/mm^2 • Coal samples were high ranked with low ash content • Among macerals, huminites

		were the major constituent
North Eastern Coals		
Singh et al [33] *(Petrographic and geochemical studies on coals in Tiru Valley in Nagaland)*	• Sampling was done using pillar coal sampling technique • Sample was reduced to 18 mesh size for petrographic analysis and 70 mesh size for proximate analysis • Leitz Orthoplan Pol microscope attached with a wild Photoautomat MPS 45 control module was used for microscopic study	• Rank of the coal varied between sub-bituminous and bituminous-D in rank • Vitrinite constitutes 98% of the total maceral content • Volatile matter in daf basis ranged from 34% to 48% in daf basis • Coal is high on sulphur content that leads to pyrite formation
Sahoo et al [34] (*Geochemical study of coal mine discharge in Jaintia Hills coalfields*)	• Water samples from mine discharge, dug wells and AMD affected streams were collected and filtered • Sample was treated with acid (4% HCl for Fe^{+2} and HNO_3 for other metals) • Acid treated sample is stored in PP bottles at 4°C • PHREEQC version 2.15 was used to model the geochemical aspects • Ion selective electrode (ORION 5 star)was used for the determination of pH	• pH of drainage from mine is found to be highly acidic (1.6-4.8) • SO_4^{2-} found in major quantities in mine discharge • In dug-wells, HCO_3^- was the most dominant radicle • Iron and aluminium were the most found dissolved metals in mine drainage
Saikia et al [35] (*Geochemical and nano - mineralogical analysis of coal and its mining by-products, in Assam*)	• Samples were taken from Ledo and Tikak collieries • Standard (ASTMD2234M, 2010) sampling method was used for coal sampling • Grab samples from the over-burden were made from the two sites. • Two fly ash samples were taken at an interval of 24 hours from an electrostatic precipitator at a PC power plant at Nagaon	• Coals had high organic sulphur content • Results from ultimate analysis reveal more than 10% carbon content in over burden • Vitrinite takes up the major portion of the total organic content of coal (95% in mmf basis) • Over burden also is found to be rich in vitrinite (10% -15% in mmf basis)

4. Conclusions

Geological parameters have significant impacts on UCG deployment in a region. A very brief review of the geological parameters in lignite-fields from Rajasthan and Gujarat, along with coalfields from North-eastern India has been carried out. Some previously published results relating to UCG deployment in Rajasthan and Gujarat indicate their high feasibility in such reasons. As regards to rock mechanics perspectives, development of new models is extremely necessary, since the present computational models developed for Indian coal mines are likely to be proven ineffective at temperatures in excess of 1200 ^{0}C. New empirical models need to supplement and improve newly initiated computational modelling techniques for Indian coalfields at such high temperatures. A framework for such a model is depicted in Figure 5 and may hold scope for future developments in UCG.

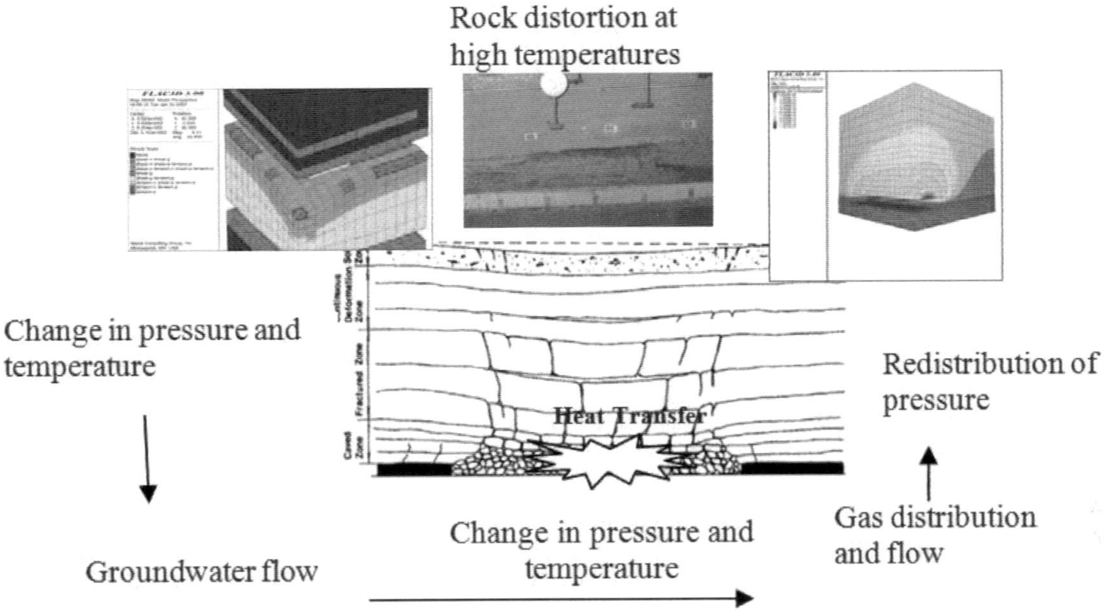

Figure 5. Framework for UCG model development

References

[1] Garg A and Shukla P R 2009 *Energy* **34** 1032-1041

[2] Chikkatur A P, Sagar A D and Sankar T L 2009 *Energy* **34** 942-953

[3] Blinderman M S, Saulov D N and Klimenko A Y 2008 *Energy* **33** 446-454

[4] GSI 2015 Inventory of Geological Resource of Indian Coal as on 01.04.2015, available at http://www.portal.gsi.gov.in/gsiDoc/pub/coal_resource_report_nenr_er.pdf

[5] Singh A K., Singh R, Maiti J, Kumar R and Mandal P K 2011 *International Journal of Rock Mechanics and Mining Sciences* **48** 805-818.

[6] Singh R and Singh T N 1999 *Geotechnical and Geological Engineering* **17** 21-35

[7] Coulthard M A 1999 *Geotechnical and Geological Engineering* **17**, 373-385

[8] Sheorey P R 1997 *Empirical Rock Failure Criteria* (A A Balakema)

[9] Sheorey P R, Loui J P, Singh K B and Singh S K 2000 *International Journal of Rock Mechanics and Mining Sciences* **37** 801-818

[10] Li H Z, Guo G L, Zha J F, Yuan Y F and Zhao B C 2016 *Bulletin of Engineering Geology and the Environment* **75** 1133-1142

[11] Laouafa F, Farret R, Vidal-Gilbert S and Kazmierczak J B 2016 *Mitigation and Adaptation Strategies for Global Change* **21** 547-576

[12] Mellors R, Yang X, White J A, Ramirez A, Wagoner J and Camp D W 2016 *Mitigation and Adaptation Strategies for Global Change* **21** 487-500

[13] Ram S, Kumar D, Singh A K, Kumar A and Singh R 2017 *International Journal of Rock Mechanics and Mining Sciences* **93** 152-162

[14] Luo J A and Wang L 2011 *Rock Mechanics and Rock Engineering* **44** 749-754

[15] Yu Y, Liang W, Hu Y and Meng Q 2012 *International Journal of Rock Mechanics and Mining*

Sciences **51** 91-96
[16] Liu S and Xu J 2014 *International Journal of Rock Mechanics and Mining Sciences* **71** 188-193
[17] Sirdesai N N, Singh R, Singh T N and Ranjith P G 2015 *Proceedings of the International Association of Hydrological Sciences* **372** 455-462
[18] Malkowski P, Kaminski P and Srzypkowski K 2012 *AGH Journal of Mining and Geoengineering* **36** 231-242
[19] Zhang W, Sun Q, Hao S, Geng J and Lv C 2016 *Applied Thermal Engineering* **98** 1297-1304
[20] Akanksha, Singh A K, Mohanty D and Jena H M 2017 *NexGen Technologies for Mining and Fuel Industries* (New Delhi: Allied Publishers) 1425-1430
[21] Sanga B, Mohanty D, Singh A K and Singh R K 2017 *NexGen Technologies for Mining and Fuel Industries* (New Delhi: Allied Publishers) 1431-1436
[22] Chaulya S K et al 2016 Development of Feasibility Assessment Model for Adaptation of Underground Coal Gasification Technology in North East Region of India (submitted to Department of Electronics & Information Technology (DeitY), Ministry of Communication and Information Technology, Government of India, New Delhi)
[23] Daggupati S, Mandapati R N, Mahajani S M, Ganesh A, Mathur D K, Sharma R K and Aghalayam P 2010 *Energy* **35** 2374-2386
[24] Prabhakar A, Sadhukhan A K, Kamila B and Gupta P 2017 *Energy & Fuels* **31** 2652-2662
[25] Singh A K, Singh M P, Sharma M and Srivastava S K 2007 *International Journal of Coal Geology* **71** 153-175
[26] Mukherjee S and Srivastava S K 2006 *Energy & Fuels* **20** 1089-1096
[27] Singh P K, Singh M P and Singh A K 2010 *International Journal of Coal Geology* **82** 1-16
[28] Singh P K, Singh M P, Singh A K, Naik A S, Singh V K, Singh V K and Rajak P K 2012 *Energy Exploration & Exploitation* **30** 131-151
[29] Rao M R, Sahni A, Rana R S and Verma P 2013 *Journal of Earth System Science* **122** 289-307
[30] Mallick M, Dutta S, Greenwood P F and Bertram N 2009 *Journal of the Geological Society of India* **74** 16-22
[31] Singh P K, Rajak P K, Singh V K, Singh M P, Naik A S and Raju S V 2016 *Energy Exploration & Exploitation* **34** 140-157
[32] Mukherjee A K, Alam M M, Mazumdar S K, Haque R and Gowrisankaran S 1992 International Journal of Coal Geology **21** 31-44
[33] Singh P K, Singh M P, Singh A K and Naik A S 2012 *Energy Exploration & Exploitation* **30** 171-191
[34] Sahoo P K, Tripathy S, Equeenuddin S M and Panigrahi M K 2012 *Journal of Geochemical Exploration* **112** 235-243
[35] Saikia B K, Ward C R, Oliveira M L, Hower J C, De Leao F, Johnston M N, O'Bryan A, Sharma A, Baruah B P and Silva L F 2015 *International Journal of Coal Geology* **137** 19-37

Acknowledgements
Authors are thankful to Dr. Pradeep K Singh, Director, CSIR-CIMFR, for his kind permission to publish this manuscript and also to CSIR Network Project *CoalGasUrja* (ESC-0302) for financial support. AKS thanks Dr. D. Mohanty, Akanksha, Bageshwari Sanga and Jayant for support, while RS thanks Arun Kumar Singh, Ashok Kumar, Rakesh Kumar and Sahendra Ram for collaboration over the years.

IOP Conf. Series: Earth and Environmental Science **76** (2017) 012003 doi:10.1088/1755-1315/76/1/012003

Global Development of Commercial Underground Coal Gasification

M S Blinderman

Ergo Exergy Technologies Inc.
Montréal, Québec, H2Y2R6 CANADA
michael.blinderman@ergoexergy.com

Abstract. Global development of Underground Coal Gasification (UCG) is considered here in light of latest trends of energy markets and environmental regulations in the countries that have been traditional proponents of UCG. The latest period of UCG development triggered by initial success of the Chinchilla UCG project (1997-2006) has been characterized by preponderance of privately and share-market funded developments. The deceleration of UCG commercialization has been in part caused by recent significant decrease of world oil, gas and coal prices. Another substantial factor was lack of necessary regulations governing extraction and conversion of coal by UCG method in the jurisdictions where the UCG projects were proposed and developed. Along with these objective causes there seem to have been more subjective and technical reasons for a slowdown or cancelation of several significant UCG projects, including low efficiency, poor environmental performance, and inability to demonstrate technology at a sufficient scale and/or at a competitive cost. Latest proposals for UCG projects are briefly reviewed.

1. Modern stage of UCG development (1997-2017)

A long period of global UCG development spawned by the energy crisis that started in1973 was completed by the Rocky Mountain 1 trial in the US in 1988 and the European UCG trial in Spain in 1992. Following several years of lull and uncertainty, the Chinchilla UCG project in Australia marked the beginning of new era of UCG development in Australia, New Zealand, South Africa, Europe, Canada and the US. Spanning almost 20 years, this latest stage of UCG development was distinguished by a preponderance of privately funded projects with a significant share of the capital raised from stock markets.

It appears that this latest stage of UCG development has suffered considerably from the drop in fossil fuel prices in the world markets, and from the global commodity market slowdown. Whereas the reduced oil and natural gas prices seem to have affected new and existing UCG projects by decreasing projected sale prices of UCG products, the corresponding precipitous drop in coal price robbed many UCG proponents of the revenue that was intended for investment into new UCG projects.

An example of the latter was the shutdown of the Huntly West UCG plant in New Zealand.

The commodity market slowdown led to the slower development the Majuba UCG project in South Africa. Reduced economic performance and the suppressed energy prices in, for instance, North American markets seem to have led to closure of the Swan Hills ISCG Project and deterred start-up of the Parkland County UCG project in Canada.

The other factor in limiting UCG activity in the world is the lack of preparedness of environmental regulations and a gross misunderstanding and misinformation on UCG within environmentally concerned communities, caused no doubt by scarcity of factual information on UCG. This regrettable state of public awareness has led to reluctance of local authorities to approve new UCG projects in several jurisdictions, in particular, in Queensland, Australia and Scotland, United Kingdom.

Along with these objective reasons for a slow-down in the global UCG development, there have been other, possibly more subjective technical grounds for a slowdown or demise of several widely anticipated UCG commercialization projects. In order to commercialize a UCG technology, its performance must usually be proven and demonstrated at a pilot scale. The pilot plants are supposed to demonstrate technical parameters of the technology, such as syngas yield per a unit of coal mass, syngas yield per a unit of injection agent flow rate, coal gas efficiency, carbon conversion efficiency, and the overall thermal efficiency of the process. Besides, pilot plant should prove environmental performance of the technology by demonstrating controlled and limited environmental impacts and the operator's ability to control and mitigate environmental risks. Pilot plant is also relied upon to show competitiveness of capital and operational costs and provide cost data for financial modeling of a full-scale, commercial plant. This presumes that the pilot plant in its design, scale and duration is representative of the process to be used at commercial scale and is capable to demonstrate the key salient features of the full-scale process in terms of its technical, environmental and financial parameters.

All UCG projects initiated since 1997 have operated as pilot plants. Four of them used the εUCG™ technology supplied by Ergo Exergy Technologies Inc and will be reviewed elsewhere. Several UCG pilot projects that have used other UCG technologies so far have not led to attaining commercialization.

2. Recent UCG pilot operations

The Swan Hills UCG trial has been conducted in Alberta, Canada in 2009-2011. It targeted a 7m thick seam of sub-bituminous coal at the depth of 1,450m. The UCG technology employed was described as linear CRIP. The underground system has been constructed by drilling directional hole with a 1,400m long in-seam section and a vertical production well [1].

The information reported by the project operators is scarce. There is no data on the quantity, quality, temperature, pressure, flow rate, liquid components, mechanical impurities of the syngas produced. Equally no data is provided for duration of gas production or parameters of injected agent such as oxygen-water ratio, flow rate and pressure. No mention has been made of the process efficiency, extraction rates, and total amount of consumed coal. Environmental performance has not been considered in the report, and no mention of environmental monitoring system has been made except for a microseismic monitoring well that seems to have been used for process monitoring.

It appears that no detailed site characterization has been undertaken prior to commencement of gasification and no obligation of monitoring the potential underground impacts has been required by the plant permit conditions. As we understand from the report, the gasification process lasted not more than several days in the row and it is unclear how many gasification attempts have been made. Some operational information can be drawn from a government regulator's incident investigation report [2]. On the sixth day of an apparent resumption of the process there was a blowout, explosion and fire in the directional injection well resulting in the loss of wellhead, release of gas and liquids into atmosphere, and fires in the nearby woods. The plant has been shut down and abandoned.

Linear CRIP system attempted in this trial, in our view, has the following inherent limitations:

1. Despite CRIP maneuvers, efficiency of gasification and quality of the syngas inevitably deteriorate with each subsequent maneuver, primarily due to the detrimental influence of multiple previously depleted CRIP reactors, downstream from the active reactor, through which the syngas must flow to reach production well.
2. One linear CRIP module trialed in Swan Hills is too small to support a commercial use of the syngas. Multiple modules would be required to feed a standard size power or chemical plant. These modules must be separated by barrier pillars to ensure independent operation

of the modules and prevent cross-flow of injection agent and produced syngas. Extraction efficiency within one linear CRIP module has been recently estimated at approximately 28% [3]. Taking into account the necessary width of the barrier pillars, the overall extraction efficiency of a mine plan consisting of multiple CRIP modules with barrier pillars would be less than 10% - far below the coal mining industry average.

3. Combined effect of syngas quality deterioration and very low extraction efficiency is a low reliability and poor use of natural resource, which in our view renders the method of linear CRIP unacceptable as a commercial coal extraction and conversion technology.

Although the project proponents seem to claim that the results of Swan Hills pilot operation may be useful in commercializing UCG at a great depth, given what is known about operational results and in absence of any news on the project for the last four years, we believe that commercialization of the process trialed at Swan Hills is unlikely.

The Chinchilla II project operated from 2007 till 2013. Unlike the Chinchilla I UCG project that was developed and operated using the εUCG™ technology, between 1997 and 2006 at the same coal lease [4], the Chinchilla II was designed, constructed, operated and managed with absolutely no involvement of Ergo Exergy Technologies Inc.

The Chinchilla II pilot plant is located near town of Chinchilla in southeastern Queensland, Australia. It targeted a 135m deep, 10m thick seam of sub-bituminous coal. Although technical information about the operations is very limited, it seems that the project utilized mostly linear CRIP method. There were four consecutive attempts at UCG at four different locations within the same property. Details of the process at any of these trials have not been made public, so there is no information on composition of injection agent (air/oxygen flow rates and pressure) nor there is any clear information on flow rates, pressures, temperatures, composition, impurities or stability of syngas production, which would not give us a chance to appraise the process efficiency. Total coal resource converted by four trials combined has not exceeded 8,000t. Starting from 2013, Queensland government announced a major investigation into environmental performance of the Chinchilla II plant. Department of Environmental and Heritage protection (QEHP) reported that in the course of this investigation

- over 230 boreholes were drilled, water and soil samples were collected from 13 surrounding farms;
- laboratory tests confirmed presence of carbon monoxide, hydrogen, hydrogen sulphide, BTEX and other chemicals in the collected samples;
- there was "scientific evidence of operation above hydrostatic pressure, fracturing the landform, and excursion of contaminants".

QEHP also alleged that plant owner failed to report numerous plant incidents, including several listed below:

- Fire that caused site evacuation in 2007
- Persistent leaks of toxic gas into air and groundwater from 2007 till 2011
- Worker's claims about their ill health as result of "uncontrolled releases" of gas at site in 2007-2013.

As a result of this investigation QEHP charged the company with irreversible damage "to more than one environmental receptor (the atmosphere, vegetation, water and soil)" and laid five criminal charges against the plant owner. A 320 km^2 exclusion zone around 1 km^2 plant was established restricting farming activities there (reportedly to be extended further). According to recent reports, five officers of the company are also on trial for similar charges [5]. In the meantime, the company that owns and operates the plant has declared bankruptcy and is in the process of liquidation.

In light of all the above we tend to believe that commercialization of the UCG process trialed in the Chinchilla II project is unlikely.

Bloodwood Creek is the name of another UCG pilot project conducted in Queensland, Australia in 2008 through 2012 [6]. It targeted a 200m deep, 9m thick sub-bituminous coal seam. Method of

IOP Conf. Series: Earth and Environmental Science **76** (2017) 012003 doi:10.1088/1755-1315/76/1/012003

parallel CRIP has been employed with 500m long in-seam panel. Two trials of gasification have been made with the first panel abandoned early in operations, apparently due to irreversible problems with injection wells, caused by explosion and blockage. Second trial reportedly produced gas of stable quality. There has been one reported instance of GW contamination in operations. Following the panels shutdown, residual contaminants reported to remain in the cavity.

Serious concerns with material balance in demonstration of parallel CRIP technology have been expressed based on CFD modeling of the gas flow within a parallel CRIP panel [7]. The modeling shows that in the conditions described above the oxygen of injected air will not react with the coal seam near injection point but rather proceed into the volume of created cavity to be consumed in parasitic reactions with syngas created previously, thereby causing deterioration of produced gas quality. This means that *raison d'être* of CRIP method – to improve stability of the process while increasing size of underground cavity – does seem unattainable in either linear or parallel CRIP configurations. Besides, all the issues with extraction efficiency discussed above for linear CRIP are equally valid for its parallel incarnation. Uncontrollable variability of gas quality and quantity together with low extraction efficiency appear to be the technical barriers that the proponents are still to overcome on the path to commercializing the CRIP-based UCG methods.

Several UCG plants, operated in the past 20 years, have been based on εUCG™ technology and are considered in a separate paper.

3. Current developments

In many parts of the word where there is no sign of pending energy abundance (including that of shale oil and gas), UCG development remains the only and imperative option for supplying affordable energy and hydrocarbon feedstock for local industrial and retail markets. Examples of such locations include South Africa, India, Pakistan, Bangladesh, and Indonesia.

It is clear that a new stage of UCG development must be based on a solid foundation of specific and comprehensive UCG regulation covering environmental protection, potential conflicts of ownership of mineral and petroleum rights, royalty regimes etc.

Many countries with large coal resources and lack of conventional oil and gas are now focusing on proposing and developing a detailed UCG regulatory framework. Among them are China, India and South Africa. These efforts are spearheaded by appropriate governmental offices, and there are indications that the regulations in these jurisdictions may become functional within the next 2-3 years.

There are a number of UCG projects that are now being prepared in anticipation of the pending regulations.

In China, there are reportedly four proposed UCG projects targeting power generation, and supply of syngas to a Fischer-Tropsch facility and a synthetic methane plant. Only scarce information has been made available by the developers.

In India, the Central Government has specified a pathway for government-owned corporations to develop UCG plants at a pilot scale. The coal and lignite blocks that would be allocated to these companies for UCG development have been identified; and the work should start in earnest once the pending regulations are available.

In South Africa, there are at least three projects that are now anticipating new UCG regulation including that of groundwater licensing. They include the Majuba UCG project development by Eskom Holdings Limited, the 50-MWe Theunissen UCG project owned by African Carbon Energy, and the Sterkfontein project developed by Oxeye Energy.

In the meantime, it appears that in several locations existing regulations could be sufficient to support a UCG project much sooner than that.

The following is a short list of the UCG projects that are currently being developed. This is not a comprehensive list, as it is rapidly evolving.

There are at least three UCG projects currently proposed in South and Central Australia, including Leigh Creek UCG project. This project is targeting a large power plant, taking advantage of the

IOP Conf. Series: Earth and Environmental Science **76** (2017) 012003 doi:10.1088/1755-1315/76/1/012003

current lack of power generation capacity in the area. The target coal deposit has been considered for UCG starting from pioneering work by Dr. Ian Stuart and Dr. Len Walker in 1982.

In Indonesia, there have been discussions of three UCG projects for which the government intends to grant operational licenses. These include a project developed by Phoenix Energy Ltd, which would produce electricity in two locations, starting from a small power plant and leading on to a larger IGCC facility with a 100-200 MW capacity.

In Alaska, a joint venture between Laurus Energy and CIRI is developing a power and synthetic methane project with ultimate capacity of 150MW and 70PJ/a, respectively. The project is in the stage of site characterization.

The Parkland project in Alberta, Canada, also owned by Laurus Energy and mothballed earlier due to low energy prices in the province, is now planned to re-start targeting a new chemical product that seems to offer better market competitiveness.

UCG projects in the UK, developed by Cluff Natural Resources and targeting offshore coal deposits have been recently put on hold in Scotland but appear to have a chance to still continue in England.

Apart from UCG projects pursuing clear commercial outcomes, there are a number of UCG projects that are conducted primarily for R & D purposes. These include the recently concluded European TOPS research project [8] that considered technology options for coupled UCG and CO_2 capture and storage; and HUGE - Hydrogen Oriented Underground Coal Gasification for Europe, development largely by Polish researchers in Główny Instytut Górnictwa. The HUGE demonstration installation was built on the premises of CCTW Mikołów in the Underground Testing Range [9].

A number of the above projects are relying on the CRIP UCG technique developed in 1975 by Lawrence Livermore National Laboratory in the USA. However, a majority of UCG projects, both the ones operated in the last 20 years and those currently under development, have been using the εUCG™ technology provided by Ergo Exergy Technologies Inc. (Canada). The latter technology also supports several new UCG projects under consideration in Turkey, India, Canada, Argentina, Pakistan and other locations.

References

[1] Swan Hills Synfuels LP 2012 *Swan Hills In-Situ Coal Gasification Technology Development – Final Outcomes Report* (Alberta, Canada)

[2] Alberta Energy Regulator 2014 *AER Investigation Report: Swan Hills Synfuels Ltd., Well Blowout, October 10, 2011* (Calgary: Alberta Energy Regulator)

[3] McVey T 2011 Final Report: Technoeconomic Evaluation of Underground Coal Gasification (UCG) for Power Generation and Synthetic Natural Gas (Livermore: Lawrence Livermore National Laboratory)

[4] Walker L, Blinderman M and Brun K 2001 *Proc. 2001 Gasification Technologies Conference*, (San Francisco, Gasification Technology Counsil).

[5] Queensland Department of Heritage and Environment Protection http://www.ehp.qld.gov.au/management/linc-energy/ (accessed April 20, 2017)

[6] Mallett C and Hains J 2015 *Proc. The Seventh Int. Conference on Clean Coal Technologies*, (Krakow, IEA)

[7] Blinderman M, Dvornikova E V, Orlov G V, Sundaram B and Klimenko A 2016 *Proc. 8th Int. Freiberg Conference on IGCC & XtL Technologies* (Cologne: Bergakademie Freiberg) p 64

[8] Durucan S, Korre A et al. 2014 *Energy Procedia* **63** 5827

[9] Mocek P, Pieszczek M, Świądrowski J, Kapusta K, Wiatowski M and Stańczyk K 2016 *Energy* **111** 313

IOP Conf. Series: Earth and Environmental Science **76** (2017) 012004 doi:10.1088/1755-1315/76/1/012004

Underground Coal Gasification – Experience of ONGC

P K Jain*

Institute of Reservoir Studies, ONGC, Ahmedabad, India

* Email: jain_pk7@ongc.co.in

Abstract. Underground Coal Gasification (UCG) is expected to be game changer for nation like ours that requires large amounts of energy but have few natural resources other than coal. ONGC, being an integrated energy company and due to synergy between E & P operations and UCG, envisaged opportunities in UCG business. Its first campaign on UCG started in 1980s. With its initiative, a National Committee for UCG was constituted with representatives from Ministry of Petroleum, Dept. of Coal, CSIR, CMPDIL, State of Gujarat and ONGC for experimenting a pilot. It was decided in mid-1986 to carry out a UCG pilot in Sobhasan area of Mehsana district which was to be funded by OIDB. Two information wells were drilled to generate geological, geophysical, geo-hydrological data and core/coal samples. 3-D seismic survey data of Mehsana area was processed and interpreted and geological model was prepared. Basic designing of pilot project, drilling and completion, strategy of process wells and designing of surface facilities were carried out. The project could not be pursued further due to escalation in cost and contractual difficulty with design consultant. ONGC second UCG campaign commenced with signing of an agreement of collaboration (AOC) with Skochinsky Institute of Mining (SIM), Russia on 25th November 2004 for Underground Coal Gasification (UCG). In parallel, MOUs were signed with major coal and power companies, namely, Gujarat Industries Power Company Ltd (GIPCL), Gujarat Mineral Development Corporation Ltd (GMDC),Coal India Ltd (CIL), Singareni Colliery Company Ltd (SCCL) and NLC India Ltd. Under the AOC, suitability study was carried out for different sites belonging to MOU companies. Only Vastan mine block, Nani Naroli, Surat, Gujarat was found to be suitable for UCG. Therefore, subsequent stages of detailed characterization & pilot layout, detailed engineering design were taken up for Vastan site. After enormous efforts for quite long since 2006, in the absence of UCG policy with Ministry of Coal (MoC), MoC finally allotted in-principle Vastan Lignite block to GIPCL in Aug. 2015. The project was to be carried out through a joint venture between ONGC and GIPCL. Unfortunately, efforts lacking sincerity were made by GIPCL for JV. MOC also did not bother adequately to monitor development of JV between ONGC and GIPCL. And finally, GIPCL citing the company to be small sized and it being without any experience on UCG, withdrew from the project in Dec. 2016. Now the block allocation process for the Vastan will have to be initiated afresh by MOC. The future of ONGC yet another UCG campaign seems to have once again hanged in balance. In view of the association with UCG for the decade and based on the feedback from the world-wide status of the technology, the author tries to make important suggestions in the paper for expeditious and efficient implementation of UCG technology in the country.

1. Introduction to UCG Process

Underground Coal Gasification (UCG) is a method for exploitation of energy from coal resources which are otherwise regarded as either uneconomical to work on by conventional underground coal extraction methods or are inaccessible due to depth, geology or other mining and safety considerations. Through UCG, coal is converted into combustible gas which can be used for industrial heating (Figure 1), power generation or the manufacture of hydrogen, synthetic gas or fuel [1]. The gas can be processed to remove its CO_2 content thereby providing a source of clean energy with minimal greenhouse gas emissions.

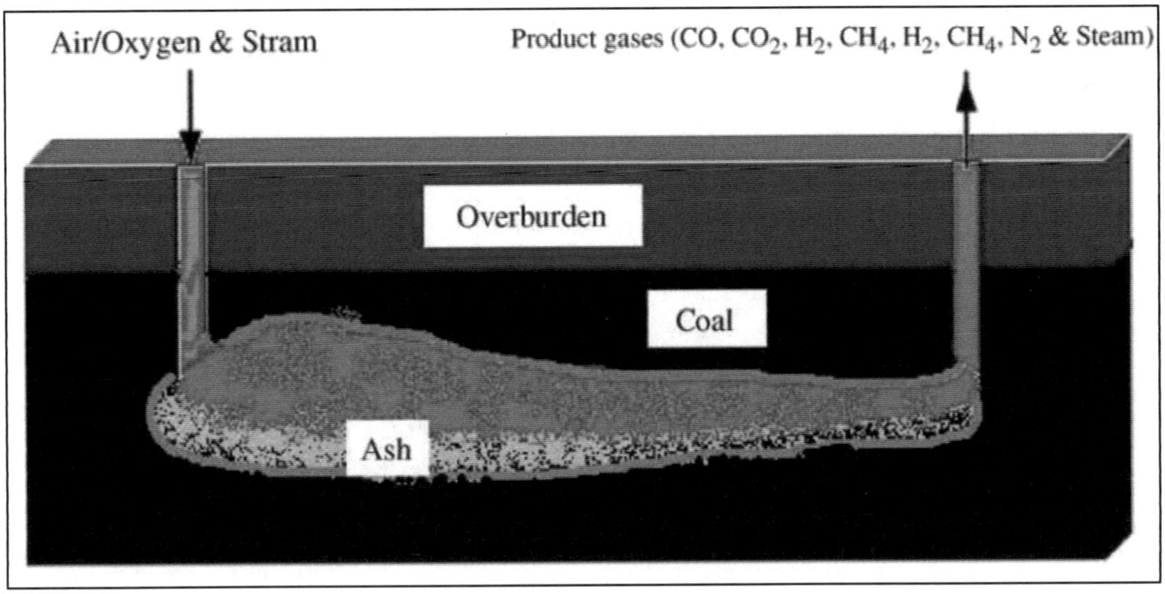

Figure 1 Schematic of UCG process [2]

The basic UCG process involves drilling a combination of two wells into coal, one for injection of the oxidants (water/air or water/oxygen) and the other, some distance away, for bringing the product gas to the surface. The end product is a combustible syngas containing hydrogen, carbon monoxide and methane along with few other gases produced as result of thermo-chemical reactions. Major reactions are as follows:

Pyrolysis

Coal + Heating (in absence of oxygen) $\rightarrow$ Char + CO_2 + CO + H_2O + H_2 + CH_4

Char Oxidation

Char + Oxygen $\rightarrow$ CO_2 + H_2O

Steam Gasification

Char + H_2O $\rightarrow$ CO + H_2

Water Gas Shift Reaction

CO + H_2O $\rightarrow$ CO_2 + H_2

Since the combustible components in syngas are mainly carbon mono-oxide (CO) and hydrogen (H_2), its calorific value (CV) is usually low (nearly $1/10^{th}$ of natural gas, NG) (with air as oxidant) and around $1/5^{th}$ of NG when only oxygen as oxidant is used.

2. Potential for UCG development in India

In India, almost 40% of the available coal resources are deep seated that is beyond 300 meters depth and mining these deposits in many cases through conventional mining techniques is difficult as well as capital intensive. The exploration activity carried out by agencies like GSI, CMPDI, MECL & NLC etc. has established a coal resource inventory of 306 billion tonnes and lignite inventory of about 44 billion tonnes. However, only less than 1/6th of the coal is economically accessible through open cast mining. Out of total coal, nearly 66% could be potential candidate for UCG. Similar is the case for lignite as well. It is pertinent to mention here ONGC has encountered vast reserves of coal during oil & gas operation in Cambay basin. Estimated reserves in north Gujarat are of the order of 63 Billion tonnes of which 37 billion tonnes were estimated to be recoverable through UCG. In addition, the discoveries in the course of oil drilling in some basins have further, revealed existence of coal/lignite deposits at greater depths. The energy from such deposits may be harnessed through UCG technology for present day use.

Most of the activities related to underground coal mining like coal stockpiling, coal washing and waste disposal and activities at user end like ash handling and disposal are eliminated in UCG. UCG has lower overall capital and operating costs. The effect of UCG on environment is not significant as the main product coming out of the process is gas and the by products are either left in the ground, or re injected into the seam. It is therefore beneficial from health and safety point of view. UCG is considered clean coal technology as it offers separation of CO_2 from syngas and its disposal is possible [2].

3. ONGC's 1st UCG Campaign

ONGC, while drilling in search of hydrocarbons, has discovered large resources of coal in the state of Gujarat. The exploitation of these coal resources was not possible due to prevailing sub-surface conditions such as depths, geology, geo-hydrology, in-situ temperature and high gas content in the coal seam, etc. Underground Coal Gasification is the only method to exploit these un-mineable coal resources. Way back in 1981, an expert committee was constituted in relevance to UCG, with the approval of inter-ministerial meeting consisting of CMRS, CFRI, Govt. of Gujarat and ONGC. The committee visited a number of countries having UCG experience viz. France, Belgium, West Germany and erstwhile USSR and recommended that UCG is the only technology for exploiting deeper coal.

Subsequently, with the initiative of ONGC, a National Committee for UCG was constituted with the representatives from Ministry of Petroleum, Dept. of Coal, CSIR, CMPDIL, State of Gujarat and ONGC for experimenting a pilot at Mehsana in North Gujarat at a depth of 700 – 1100 m. Estimated reserves in North Gujarat are of the order of 63 Billion tonnes of which 37 Billion tonnes were estimated to be recoverable through UCG. With the concerted efforts of the Committee members it was decided in mid-1986 to carry out a UCG pilot experiment in Sobhasan area of Mehsana district which was to be funded by OIDB.

Major works carried out were:

a) Drilled two information wells UCG-1 and UCG-3.
b) All geological, geophysical, geo-hydrological data and core/coal samples collected.
c) Detailed lab studies on core and coal samples of well UCG-1, was carried out.
d) 3-D seismic survey data of Mehsana area was processed and interpreted.
e) Geological Model prepared.
f) Basic designing of pilot project, drilling and completion, strategy of process wells and designing of surface facilities.

The stumbling block for the progress was non finalization of the contract with M/s PDIL for surface engineering and design and finalization of the foreign back up consultant M/s BME through M/s PDIL. M/s BME had refused to accept the task of back up consultancy. The revised estimates were prepared under the guidelines from National committee by the UCG subcommittee. Depending on the year of completion (1992 – 1996) of the project the estimated cost of the pilot can vary in the range Rs.40 – Rs. 62 crores of which the foreign exchange components will be of the order of Rs. 16 to Rs. 26 crores (Rupees equivalent). This is against an initial estimate of the project cost of Rs. 9.6 crores. The committee decided that this R&D project should appropriately be funded by OIDB and hence OIDB was approached to provide the grant- in-aid to meet the cost of the project. The case was processed for the approval of OIDB for funding this project by the OIDB s grant- in-aid. The OIDB funding was not forthcoming due to the escalation of costs. For these reasons this project was discontinued in 1992.

4. ONGC's 2nd UCG Campaign

ONGC signed an Agreement of Collaboration (AOC) with Skochinsky Institute of Mining (SIM), Russia in November 2004 for Underground Coal Gasification (UCG). MoUs were signed with companies like CIL, SCCL, NLC, Gujarat Mineral Development Corporation (GMDC) and Gujarat Industries Power Corporation Ltd (GIPCL). Eleven sites were evaluated for suitability to UCG and out of those, only Vastan mine block, Nani Naroli, Surat, Gujarat was found to be suitable. Subsequent stages of detailed characterization & pilot layout, detailed engineering design were taken up for Vastan. Additional geological and hydro-geological data was generated for the design and execution of UCG pilot. High Resolution Shallow Seismic Survey was carried out for detailed characterization of the field. The data was analyzed and location and layout of UCG Pilot finalized. The detailed engineering design for construction of UCG pilot module is available. Ministry of Environment and Forest, Government of India has given the Environmental clearance for testing of pilot. AOC with SIM and MOUs with GIPCL, GMDC and NLC are valid at present also.

Allocation of Vastan block for UCG had consistently been followed up with Ministry of Coal (MoC) for years since 2006. Finally, MoC allotted in-principle Vastan Lignite Block to GIPCL in August 2015. The project was to be carried out through a joint venture between ONGC and GIPCL. Allottee (GIPCL) was required to enter into an agreement containing the terms & agreement of allotment namely, Coal Mine Development & Production Agreement (CMDPA) with the Central Government.

Unfortunately, efforts lacking sincerity were made by GIPCL for JV. MOC also did not bother adequately to monitor development of JV between ONGC and GIPCL. And finally, GIPCL citing the company to be small sized and it being without any experience on UCG, withdrew from the project in Dec. 2016. They are in the process of intimating MoC in this direction. Now the block allocation process for the Vastan will have to be initiated afresh by MOC. The future of ONGC yet another UCG campaign seems to have once again hanged in balance. After the block allocation, the activities required before ground work at site includes mine plan approval, grant of mining lease by state govt., land acquisition followed by approach road and site preparation.

The very casual approach on UCG by stakeholders may be attributed to (i) much delayed approval of the UCG policy framework by Govt. of India and absence of model agreement which is to be signed between government and Allottee. A policy framework has been approved by the government in Dec. 2015. According to this, a model agreement similar to CBM, i.e. revenue sharing will be followed for UCG. CMPDIL has been made nodal agency for UCG implementation. (ii) Also not available is the bid documents and evaluation criteria for allotment of UCG blocks by MoC. Both the documents; model agreement and bid-documents, are under development at present by CMPDIL, Ranchi by hiring consultant through international bidding. CRISIL is the consultant which has been awarded this work. (iii) Worldwide sluggishness on UCG commercialization. However, the countries with huge resources of un-mineable or un-economically mineable coal/ lignite need to be pro-active on usage of UCG.

5. Recommendations for effective UCG deployment in India

In view of the association with UCG for the decade and based on the feedback from the world-wide status of the technology, the author tries to make important suggestions in the paper for expeditious and efficient implementation of UCG technology in the country.

5.1. Pilot trials are necessary for gaining confidence

In the present scenario, the process of moving ahead for promotion of the technology has turned very slow. ONGC has already carried out major works on the Vastan UCG project. Site suitability, detailed site characterization including seismic survey of the lignite seams and surface facilities design have already been completed by 2009 with consultancy of Russian Expert Skochinsky Institute of Mining (SIM) [3].

The whole process as detailed above is likely to take very long time and work already done by ONGC may go waste. In view of this, it is pertinent for MoC to permit ONGC or any other collaboration to carry out only the pilot at Vastan lignite block as soon as possible without waiting for any long term policy framework and CMDPA etc.

The UCG technology is site specific and its technical success need to be first established in the country through pilots only. The economics of an UCG project are clearly dependent upon the characteristics of the location. Additionally, the concept apparently does not lend itself to a once and for all solution. The solutions must be tailored to match the site. UCG undoubtedly will not be economical for all coal resources that are slated for development. R & D is needed, in fact, to define the resources, whose UCG would be cost effective. The technology requires detailed investigation, pilot experimentation and data evaluation before embarking on full scale commercial projects. The successful R & D program in UCG would be considered viable commercialization of clean fuels from coal at prices which would be competitive with existing energy sources. The countries: South Africa, Canada, China, Australia, New Zealand have gained rich experience through pilots/ demo usage of UCG syngas during 2007-13 [4].

'Road Map to Develop UCG Technology in India', prepared by the Working Group and the R & D Committee on UCG, under the auspices of the Office of the Principal Scientific Adviser To the Government of India, (PSA, GoI), Nov, 2008, has also recommended on the above lines: The Working Group to prepare the Roadmap was constituted by R & D Committee on UCG under the Chairmanship of Shri A.K.Hazarika, Director (Onshore), ONGC Ltd. Other members of the Group were CIL, GAIL, NLC, BHEL, CMRI, ISM, ONGC, Scientist E, Office PSA-GoI, MoC, CMPDIL, RIL, Essar Oil Ltd.

As per the Road-map, 'there is an urgent need to carry out at least one UCG pilot scale study each in coal and lignite deposits, so that efficacy of the suitable technology under Indian conditions can be established [5]. This will give confidence to the Indian operators to adopt that technology and speed up the start of commercial operations. The committee was set-up to give appropriate recommendations regarding regulatory framework and related issues also. It proposed a framework to facilitate operations by various interested parties. Based on the data generated from the pilot scale studies/ commercial stage operations, this interim regulatory framework would be reviewed periodically by a committee appointed by MoC with representation from MoPNG. For this purpose, a facilitating and monitoring mechanism or committee was proposed to be set up at MoC, which was suggested as administrative ministry for UCG. A very important observation on applicability of existing act/ rules was made by the committee: Since syngas produced from UCG operations neither fall under the definition of 'Coal' nor by the definition of 'Petroleum' as per the existing regulations, a separate set of Regulations will have to be developed in due course of time for UCG operations.

As per the reports available in the public domain, China, Indonesia, Hungary, Canada and South Africa are in the process of preparing regulations on UCG in their respective countries to promote UCG. One Australian Pvt. Company, namely Carbon Energy, is participating in UCG research center in China in conjunction with China University of Mining and Technology (CUMT), one of the largest mining universities in the world.

5.2. Core Group needs to be developed

UCG operations are quite different than normal E&P operations. There is need for development of core group on UCG in any organization to take up the operational challenges. It is prudent to be technically conversant with intricacies of the UCG process. The major issues which need exposures are as follows:

- Drilling of UCG Gasifier [6]
- Cementation /Metallurgy for higher temperature – The expected temperature down hole in the cavity may vary 1000 to 1300 °C depending upon moisture contain in the lignite seam. The

completions are to be suitable for this temperature range. Additionally the cement has to be high temperature resistant.

- Suitability of site for UCG, estimation of gasifiable coal reserves and techno-commercial evaluation.
- Design of process layout
- Processing of UCG raw syngas – The syngas may contain suspended solids, H_2S, moisture nitrogen & condensate. This gas needs to be processed to make it suitable for sale.
- Handling of UCG syngas and condensate – As the constituents of the both are toxic in nature.
- Processing of condensate to extract valuable components such as phenols etc.
- Safety guideline to be observed at UCG Installation.
- Carbon capture and Disposal – UCG syngas is expected to contain lot of CO_2 which needs to be separated and disposed of. One has to be acquainted with latest development on carbon capture and sequestration (CCS) [7].

5.3. Integration with some end-usage

It is now established that well drilling and completion cost take more than 65-80% of the total cost in UCG, efforts need to make to get steady syngas volume maximizing volume with minimum number of wells. Syngas production alone may not be economically viable, it needs to be integrated with some end usage such as methanol, SNG, power, GTL, fuel cell, etc [8].

5.4. Formation of UCG Association

To promote UCG in the country, some sort of association by the energy PSUs may be formed which carryout advanced R&D in the area of UCG on the lines of South Africa and China. Recognizing the potential of UCG, Exxaro, Eskom, Sasol and Africary in South Africa have founded the South African UCG Association to champion the technology and create industry standards and training schemes.

It is with this objective that this project and the association thus formed will propel our energy starved country to the forefront of a global industry based on an unconventional technology.

6. Conclusion

The utilization of UCG technology in Indian scenario can be materialized in harnessing the energy from un-minable / uneconomically minable coal / lignite. ONGC has made two serious campaigns for the implementation of technology in the country, first in 1980's and second in first decade of current century. The implementation did not reach logical conclusion because of absence of UCG policy for so long. Now after the approval of policy frame work in December 2015 and proactive approach of the central government in developing related documents, one can hope for change in mindset of stake holders towards UCG technology and visible UCG implementation on ground.

References

[1] Gregg D W and Edgar T F 1978 *AIChE Journal* **24** 753-781
[2] Khadse A, Qayyumi M, Mahajani S and Aghalayam P 2007 *Energy* **32** 2061-2071
[3] Patro D N, Sharma D K, Singh R P, Vishwanathan, Sharma R K 2008 *7th International Conference & Exposition on Petroleum Geophysics* (Hyderabad) P-383
[4] Yang L, Liang J and Yu L 2003 *Energy* **28** 1445-1460
[5] PSA 2008 Road Map to Develop Underground Coal Gasification Technology in India (New Delhi: Principal Scientific Advisor to the Government of India)
[6] Stephens D R 1979 *Proc. 5th Underground Coal Conversion Symposium* (Alexandria, Virginia) p 163-176
[7] Roddy D J and Younger P L 2010 *Energy & Environmental Science* **3** 400-407
[8] Prabu V and Jayanti S 2012 *International Journal of Hydrogen Energy* **37** 1677-1688

IOP Conf. Series: Earth and Environmental Science **76** (2017) 012005 doi:10.1088/1755-1315/76/1/012005

Process modelling and simulation of underground coal gasification: A Review of work done at IIT Bombay

Surabhi Sharma and **Sanjay M Mahajani***

[1]Department of Chemical Engineering, IIT Bombay

* Corresponding Author Email: sanjaym@iitb.ac.in

Abstract. This paper presents the summary of the work performed over the last decade, at IIT Bombay by the UCG group. The overall objective is to determine the feasibility of a given coal for underground coal gasification and then determine the capacity of a single pair of well through modelling and simulation. It would help one to design a UCG facility for the desired rate of gas production. The simulator developed in this study seeks inputs on four important aspects: Kinetics of all the reactions under the conditions of interest, heat and mass transfer limitations, if any, the flow patterns inside the cavity and lastly the thermo-mechanical failure of the coal. Each of them requires detailed studies in laboratory. Indian Lignite from one of the reserves was chosen as a case study.

1. Introduction

Underground coal gasification is an in-situ process of synthesis gas generation which utilizes deep coal reserves (Figure 1) that are otherwise difficult to mine. India has many such coal reserves and a comprehensive review on the same has been published elsewhere [1]. In UCG process simultaneous combustion and gasification of coal are performed in the presence of gases (oxygen, steam) to generate syngas which finds its application in power generation, nitrogenous fertilizers, petrochemicals, liquid fuels and many such industries. The operation involves drilling of injection and production wells at an appropriate distance. The reagent gases are injected through the injection well into the coal seam. A cavity is formed underground and various gas phase and gas-solid reactions take place. Thus UCG process can be divided into four main steps namely: Drilling, linking, ignition or combustion and gasification.

UCG cavity is a 3-D reactor whose geometry is irregular and is associated with highly non-ideal flow patterns and also the strong interaction between several other chemical and physical processes like chemical reactions, heat and mass transport phenomena and spalling. Water influx further increases the complexity of the process. UCG is a site-specific process and the scope for adjusting the parameters is limited. Thus the development of predictive tools which can determine the behaviour of a UCG reactor under different physical and operating conditions and thus help in control and optimization of the process becomes a need of the day.

This paper summarizes the major contributions and significance of experimental and simulation studies performed at IIT Bombay in collaboration with ONGC over the last decade to develop model which predicts the performance of a UCG process by taking into account the various phenomena which affect the process thus influencing the quality and quantity of product gas [1-11]. Indian Lignite from one of the reserves was chosen as a case study.

IOP Conf. Series: Earth and Environmental Science **76** (2017) 012005 doi:10.1088/1755-1315/76/1/012005

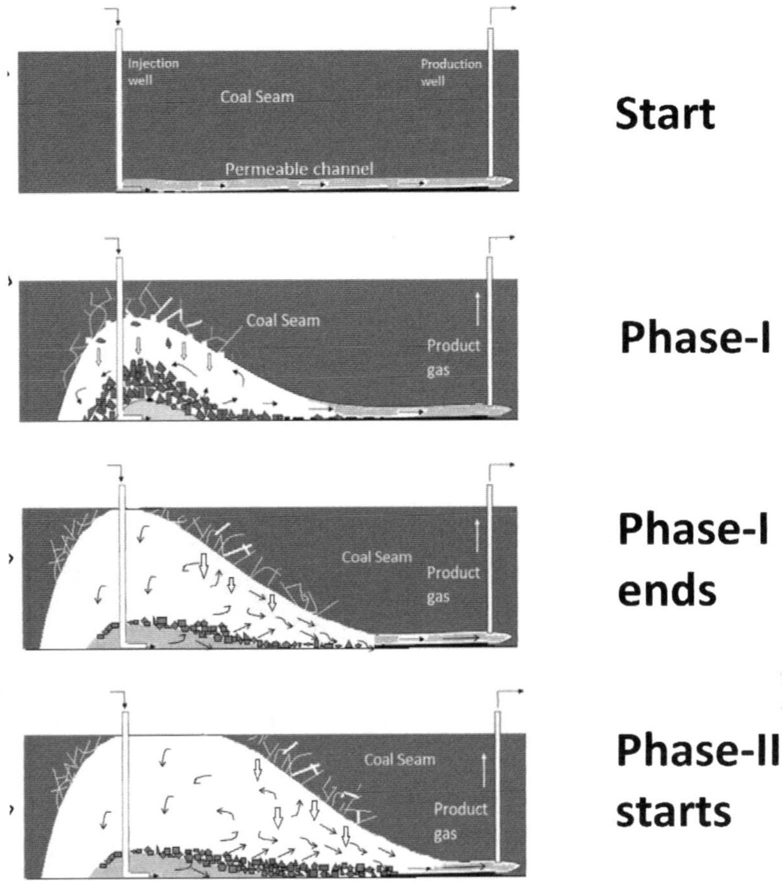

Figure 1. Underground Coal Gasification Process [2]

2. Methodology for model development

The input and output information associated with the simulator developed is shown in figure 2. Model inputs associated with kinetics, heat/mass transfer and spalling have been determined for the coal of interest.

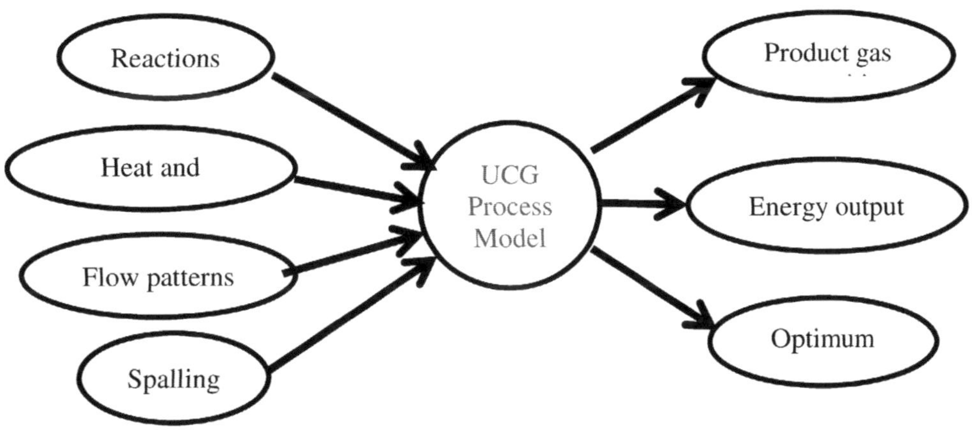

Figure 2. Input and output information for process model [10]

IOP Conf. Series: Earth and Environmental Science **76** (2017) 012005 doi:10.1088/1755-1315/76/1/012005

To determine reaction kinetics, laboratory data was generated under UCG-like conditions. Similarly, correlations for heat and mass-transfer coefficients and appropriate spalling (thermo-mechanical failure) parameters were determined for flow and thermal conditions in the UCG geometry [11]. Unlike other parameters, spalling characteristics are expected to be scale-dependent, and therefore these parameters might require tweaking for simulation of field scale experiments. To study the importance of flow fields inside the UCG cavity computational fluid dynamics (CFD) simulations were performed. Based on the detailed study of each one of these inputs the model is developed and the performance is simulated.

2.1 Reaction Kinetics

The steps involved in conversions taking places in UCG cavity are drying, pyrolysis and chemical reactions like combustion, gasification (table 1) and the kinetics of these reactions should be experimentally determined [4, 5].

Table 1. Main Reaction involved in UCG process

Reaction name	Reaction
Drying	$H_2O_{(l)} \overset{heat}{\Longrightarrow} H_2O_{(v)}$
Pyrolysis	$CH_aO_b \overset{k_{f,p}}{\Longrightarrow} a_1C + a_2CH_4 + a_3H_2O + a_4CO + a_5H_2 + a_6CO_2 + a_7C_9H_c$
Char combustion	$C + O_2 \overset{k_{f,1}}{\Longrightarrow} CO_2$
Steam gasification	$C + H_2O \overset{k_{f,2}}{\Longrightarrow} CO + H_2$
Boudard reaction	$C + CO_2 \overset{k_{f,3}}{\Longrightarrow} 2CO$
Methanation	$C + 2H_2 \overset{k_{f,4}}{\Longrightarrow} CH_4$
Water-gas-shift	$CO + H_2O \overset{k_{f,5}}{\Longleftrightarrow} CO_2 + H_2$
Gas-phase oxidation	$H_2 + \frac{1}{2}O_2 \overset{k_{f,6}}{\Longrightarrow} H_2O$,
	$CO + \frac{1}{2}O_2 \overset{k_{f,7}}{\Longrightarrow} CO_2$,
	$CH_4 + 2O_2 \overset{k_{f,8}}{\Longrightarrow} CO_2 + 2H_2O$
Tar-reforming	$C_9H_c + 9H_2O \rightarrow 9CO + \left(9 + \frac{c}{2}\right) H_2$

Kinetics of char combustion cannot be determined from usual thermo-gravimetric analysis because it is a fast reaction and also the size of coal seam is very large as compared to the size of sample used in TGA. Hence, to overcome this problem a novel reactor setup to determine the fast reaction kinetics was designed which was then used to formulate a kinetic sub-model. Kinetics of other heterogeneous reaction like steam gasification and CO_2 gasification was also studied using specifically designed TGA. To model gas-solid non catalytic reactions a new expression for effectiveness factor in case of random pore model was proposed as given in equation 2.1.4. The rate of reaction by random pore model is given by:

$$\text{rate} = k * C_{gas,S} * C_{Char,0} * (1 - X) * \sqrt{1 - \psi * \log(1 - X)} \quad (1)$$

Where,

ψ = structural parameter in random pore model

X = local char conversion

The time scales for diffusion and for changes in surface area and effective diffusivity are much different. This constancy of the surface area and that of effective diffusivity are thus considered to get an expression for Thiele modulus. Later the gas balance equations are solved taking the surface area and effective diffusivity as a functions of solid concentration.

$$M_T = \frac{d}{6} * \sqrt{\frac{k * C_{Char,0}}{D}} \tag{2}$$

$$\eta = \frac{1}{M_T}\left(\frac{1}{\tanh(3M_T)} - \frac{1}{3\,M_T}\right) \tag{3}$$

$$rate = \eta * k * C_{gas,S} * C_{Char,0} * (1 - X) * \sqrt{1 - \psi * \log(1 - X)} \tag{4}$$

This methodology is validated by comparing results with complete diffusion-reaction modeling on a spherical particle [10].

2.2. Flow Patterns

Virtual studies of residence time distribution were performed [2, 6] based on CFD for characterization of non-ideal flow patterns in irregular UCG cavity (figure 3). It may be noted that the basis of the cavity shape comes from the visual observations of cavity formed in several lab scale UCG experiments performed over a small coal block [6].

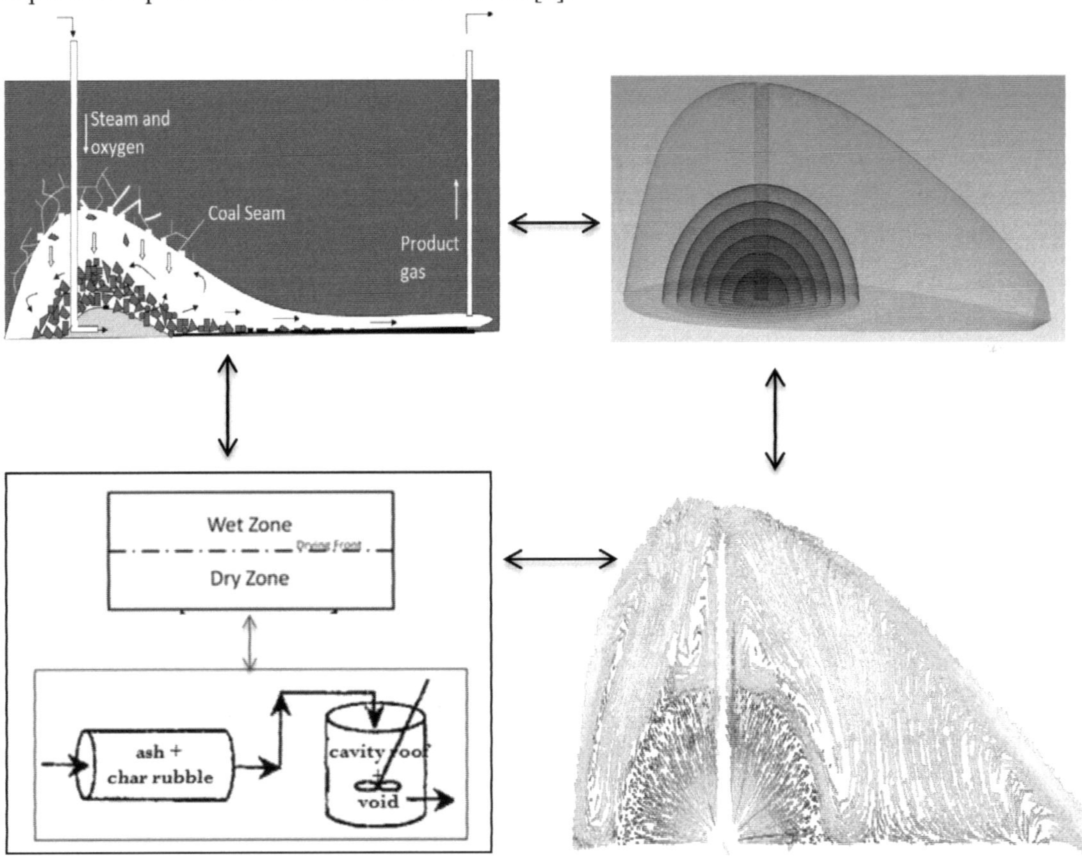

Figure 3. Virtual RTD studies to determine flow patterns and proposed compartment model [10]

The determination of flow patterns in various conditions reveals that one can approximate the cavity as a simplified compartment model that includes Plug flow reactor (corresponding to the rubble fallen on the ground) and the CSTR for the gas flow in the void space of the cavity(Phase-I). Outflow channel may be assumed as a series of CSTRs (phase-II.). Phase-I concentrates on the vertical growth of cavity till it reaches the overburden after which starts the phase-II where horizontal growth of cavity occurs till it reaches the production well. Compartment modelling simplified the computational intensiveness many fold and made predictions faster. It may be noted that the basis for the shape of cavity comes from several laboratory experiments performed on smaller coal blocks. For details, one can refer the work by Daggupati et al. [6].

2.3. Heat and Mass transfer

The various heat and mass transfer interactions considered during modelling are shown in figure 4. Different modes of heat transfer viz radiation, convection and conduction were considered and appropriate correlations for heat and mass transfer were used in energy and mass balance equations for both gas and solid species in different zones [11].

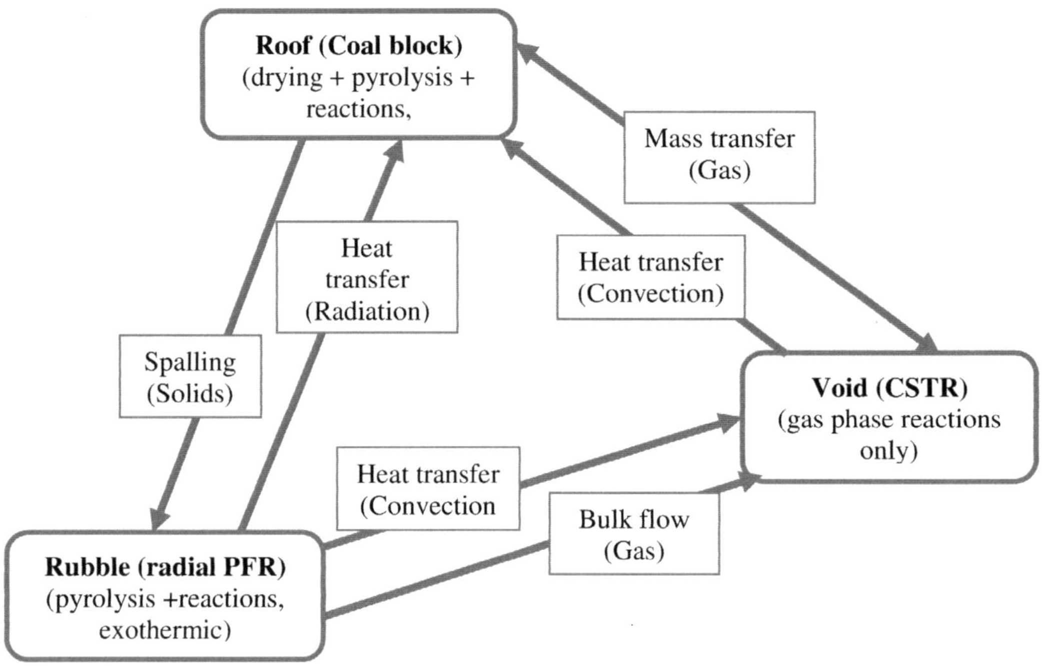

Figure 4. Heat and mass transfer interactions in UCG cavity [2]

2.4. Thermo-mechanical failure (Spalling)

Here, we determined the rate at which the coal/char particles detached from the roof of the cavity and fell on the floor. Rate of spalling is an important parameter because spalling is responsible for creating large external surface for the important gas-solid reactions taking place in the cavity and outflow channel. Thus higher the rate of spalling, higher will be the overall rate.

A special apparatus is designed and fabricated to measure the rate of spalling for the given coal (figure 5). The main part of the apparatus was a furnace in which UCG like conditions were replicated with a plate type heater at the bottom surface of its heating chamber. The heating chamber has inlet and outlet ports for the gas. A coal block holder made up of stainless steel (SS-410) was placed in the chamber, with its bottom face exposed to the hot flowing gas and thermal insulation on all other sides.

This holder was hung from a load cell that was placed outside the furnace, to monitor the weight of the coal block. Figure 6 shows the results obtained from this experiment for the amount of spalled coal with time [8, 9].

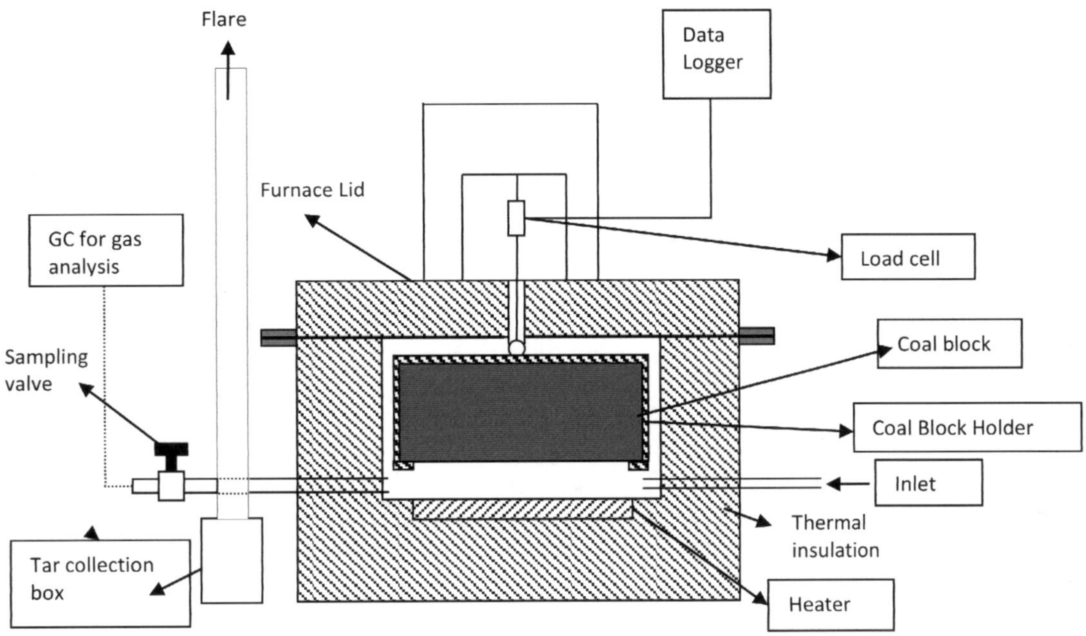

Figure 5. Schematic of the apparatus to measure spalling rate [9]

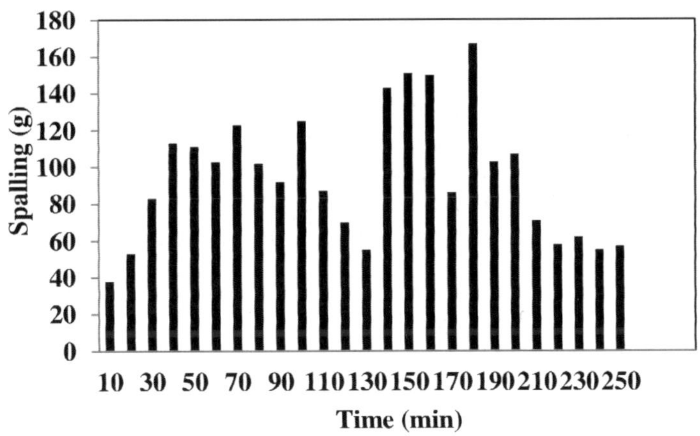

Figure 6. Amount of spalled coal particle with time [9]

It was also observed that the gas calorific value increases with spalling till a critical spalling rate but after that there is no effect. It is due to the reason that after critical spalling the surface available for reaction is large enough and the availability of reacting gas species becomes the limiting factor. The lignite coal of interest gives sufficiently high spalling rates so gas calorific value and exit gas composition were observed to be not so sensitive to a further change in the spalling rate. However, in other cases, where spalling rates are less, a significant impact on the product gas quality can be observed and it may control the performance of the system.

3. Process Model and simulation results

All the above inputs are provided to the process model, which as mentioned above, conceptualizes UCG as a set of multiple zones and compartments linked to each other through heat and mass streams. The model is a set of differential algebraic equations in time and space, and solved using in-house MATLAB program. The dynamic simulation of the model was performed varying the volume of reactor as a function of time based on the amount of coal consumed. With every spalling instance, the volume of rubble zone increases [2, 3].

Figure 7 shows temperature profiles in rubble zone at different times. After every spalling instance temperature in the rubble and roof zone changed due to relatively colder roof. Figure 8 shows plot of char density in the rubble with time.

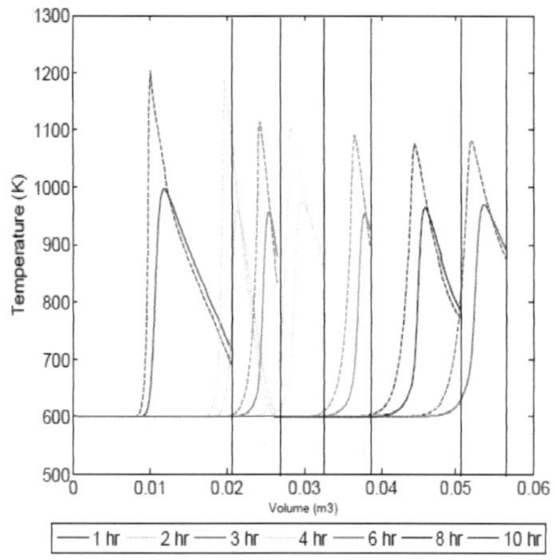

Figure 7. Temperature Profiles at different time [2]

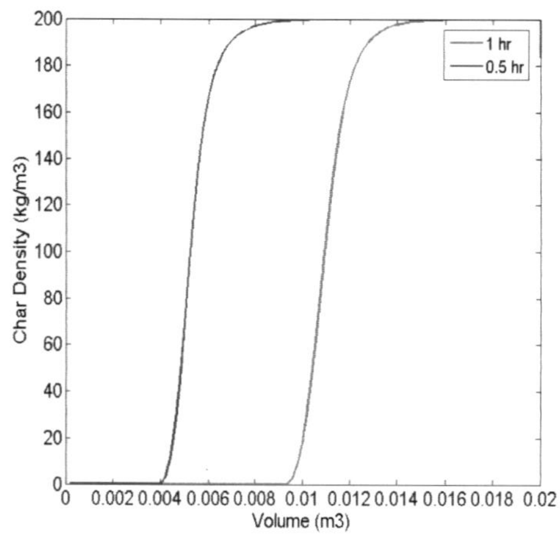

Figure 8. Char density at different times [2]

The change in calorific value with time as observed in the results is shown in figure 9 which shows change in calorific value at every spalling instance. Variation in exit gas mole fraction with time is shown in figure 10 which shows that between two spalling incidents the exit gas fractions remain in a fixed range. The results are validated experimentally with the help of laboratory-scale UCG experiments and the agreement has been satisfactory as shown in figure 11 and table 2. The simulator is computationally intensive and takes long time (20-30 hrs.) to converge to the solution. It is our future task to simplify the model without compromising on the quality of the results.

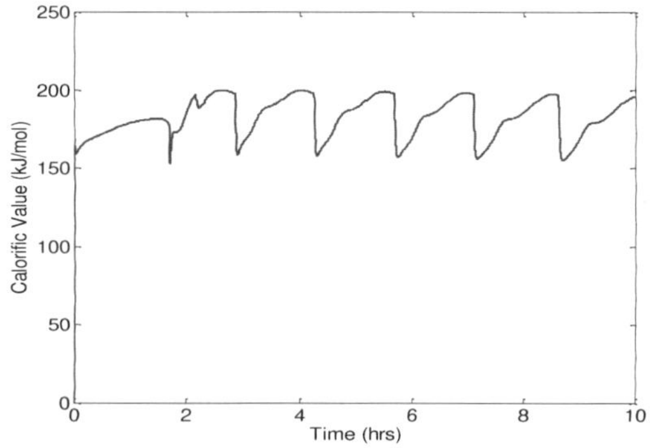

Figure 9. Calorific value of exit gas with time [2]

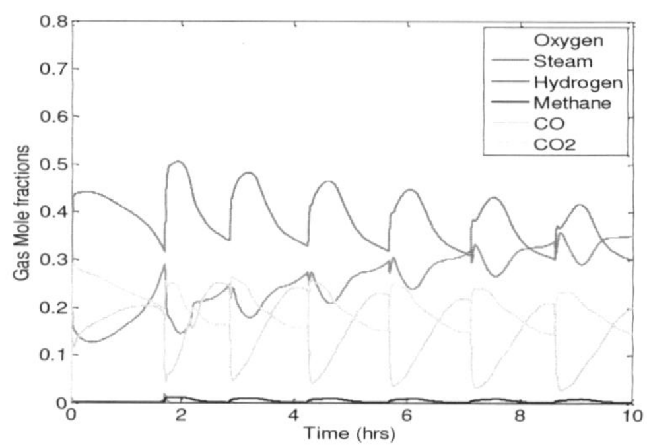

Figure 10. Exit-gas mole fractions with time [2]

Table 2. Comparison of the predicted cavity growth rate and the observed at lab-scale UCG [10]

Parameter	Lab-scale UCG	Process Model
Calorific value (kJ/mol)	~160	157
O_2 Flow Rate (kmole/min)	3.2×10^{-5}	3×10^{-5}
Total Volume Consumed (cc)	1000	1120

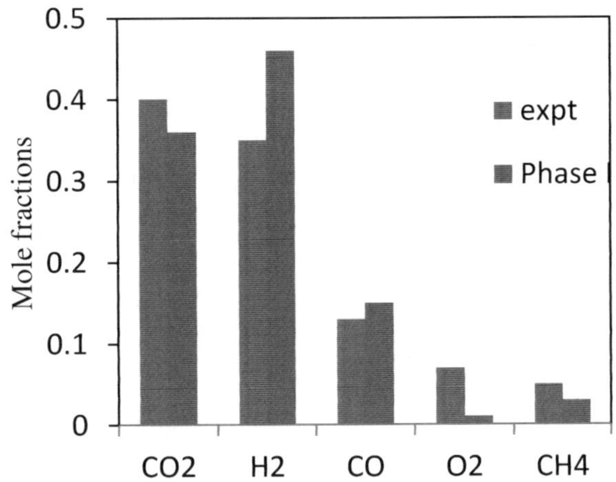

Figure 11. Comparison of experimental and simulation results for UCG results at steady state [10]

4. Conclusion

UCG process model needs inputs on kinetics, flow patterns, spalling and heat and mass transport behaviour of UCG cavity and outflow channel. The kinetic modeling is done for the heterogeneous reactions involved in the UCG process. To determine the kinetics of coal combustion, a novel reactor setup has been designed. Also a simplified procedure for determination of effectiveness factors for RPM kinetics is developed for the first time.

CFD studies were performed to characterize non-ideal flow patterns inside UCG cavity. Based on the flow patterns inside the cavity, compartment modelling approach was devised for different phases of growth of UCG cavity and the resulting model equations were solved for different zones. The attractiveness of this approach is in reducing computational burden and therefore providing faster prediction of the process performance. Spalling is an important phenomena and is responsible for generating large surface area to increase the reaction rates. The data on spalling is generated in laboratory and spalling rate was used as an input parameter. A critical spalling rate is the one beyond which the performance is insensitive to further changes.

The work provides a complete modelling solution for the UCG process. All the important chemical and physical processes have been considered in the model along with the phenomena of spalling and the model agreed well with the CFD and experimental studies performed. Efforts are being made to develop a simplified version of this model in order to reduce the computational time without compromising much on the quality of predictions.

References

[1] Khadse A, Qayyumi M, Mahajani S and Aghalayam P 2007 *Energy* **32** 2061–2071.
[2] Samdani G, Aghalayam P, Ganesh A, Sapru R K and Lohar B L 2016 *Fuel* **181** 690-703
[3] Samdani G, Aghalayam P, Ganesh A, Sapru R K and Lohar B L 2016 *Fuel* **181** 587-599
[4] Mandapati R N 2015 *Kinetic studies in Underground coal gasification* Ph.D. thesis (Indian Institute of Technology Bombay)

[5] Mandapati R N, Naidu R, Daggupati S, Mahajani S, Aghalayam P, Sapru R K, Sharma R K and Ganesh A 2012 *Industrial & Engineering Chemistry Research* **51** 15041–15052
[6] Daggupati S, Mandapati R N, Mahajani S M, Ganesh A, Mathur D K, Sharma R K, and Aghalayam P 2010 *Energy* **35** 2374-2386
[7] Daggupati S, Mandapati R N, Mahajani S M, Ganesh A, Pal A K, Sharma R K, and Aghalayam P 2011 *Industrial & Engineering Chemistry Research* **50** 277-290.
[8] Bhaskaran S, Ganesh A, Mahajani S M, Aghalayam P, Sapru R K and Mathur D K 2013 *Fuel* **113** 837-843
[9] Bhaskaran S, Samdani G, Aghalayam P, Ganesh A, Singh R P, Sapru R K, Jain P K and Mahajani S M 2015 *Fuel* **154** 326-337
[10] Samdani G 2015 *Underground coal gasification: Kinetic studies and process model* Ph.D. thesis (Indian Institute of Technology Bombay)
[11] Daggupati S 2010 *Underground coal gasification: Experimental studies and computational flow modelling* Ph.D. thesis (Indian Institute of Technology Bombay)

Acknowledgement
We acknowledge an equal contribution of Profs. Anuradda Ganesh and Preeti Aghalayam in this work and the efforts of more than 15 graduate students at various phases of this project. Furthermore the financial support by ONGC is gratefully acknowledged.

IOP Conf. Series: Earth and Environmental Science **76** (2017) 012006 doi:10.1088/1755-1315/76/1/012006

Mathematical Modelling of Coal Gasification Processes

T Sundararajan*, V Raghavan, A Ajilkumar, K Vijay Kumar

Department of Mechanical Engineering, Indian Institute of Technology Madras, Chennai, India

* Corresponding Author. Email: tsundar@iitm.ac.in

Abstract. Coal is by far the most commonly employed fuel for electrical power generation around the world. While combustion could be the route for coal utilization for high grade coals, gasification becomes the preferred process for low grade coals having higher composition of volatiles or ash. Indian coals suffer from high ash content- nearly 50% by weight in some cases. Instead of transporting such high ash coals, it is more energy efficient to gasify the coal and transport the product syngas. Integrated Gasification Combined Cycle (IGCC) plants and Underground Gasification of coal have become attractive technologies for the best utilization of high ash coals. Gasification could be achieved in fixed beds, fluidized beds and entrained beds; faster rates of gasification are possible in fluidized beds and entrained flow systems, because of the small particle sizes and higher gas velocities. The media employed for gasification could involve air/oxygen and steam. Use of oxygen will yield relatively higher calorific value syngas because of the absence of nitrogen. Sequestration of the carbon dioxide after the combustion of the syngas is also easier, if oxygen is used for gasification. Addition of steam can increase hydrogen yield in the syngas and thereby increase the calorific value also. Gasification in the presence of suitable catalysts can increase the composition of methane in the product gas. Several competing heterogenous and homogenous reactions occur during coal major heterogenous reaction pathways, while interactions between carbon monoxide, oxygen, hydrogen, water vapour, methane and carbon dioxide result in several simultaneous gas-phase (homogenous) reactions. The overall product composition of the coal gasification process depends on the input reactant composition, particle size and type of gasifier, and pressure and temperature of the gasifier. The use of catalysts can also selectively change the product composition. At IIT Madras, over the last one decade, both experimental and modelling work has been undertaken to investigate the gasification characteristics of high ash Indian coals and compare the yield with those of high grade Australian and Japanese coals. A 20 kW capacity entrained flow gasifier has been constructed and the gasification characteristics have been studied for Indian coals for different particle sizes, system pressures and air flow rates. The theoretical model incorporates the effects of Knudsen diffusion, devolatilization and various heterogenous and homogenous kinetic steps as well as two-phase flow interactions involving the gaseous and particle phases. Output parameters such as carbon conversion, cold gas efficiency and syngas composition have been compared for different grades of coals under a wide range of operating conditions. The model developed for the entrained flow gasifier predicts the gasification characteristics of both Indian and foreign coals well. Apart from the entrained flow gasifier, a bubbling bed gasifier of 100 kW capacity has also been studied. A pilot plant for the gasification of Indian coals has been set up for this capacity and its performance has been investigated experimentally as well as theoretically at different air and steam flow rates. Carbon conversion efficiency of more than 80% has been achieved.

IOP Conf. Series: Earth and Environmental Science **76** (2017) 012006 doi:10.1088/1755-1315/76/1/012006

1. Introduction

Coal is by far the major source of energy world-wide and it is likely to remain so, for some more decades to come. India is endowed with large reserves of coal which are expected to last for at least two more centuries, but the coal contains a high percentage (~40% by mass) of ash. The high ash content results in low calorific value and it is also an additional burden when coal is transported to different parts of the country from the mines. Moreover, the presence of fly-ash in the flue gas from coal combustion is a pollution hazard and additional devices such as electrostatic precipitators need to be fitted for the removal of ash particles before the flue gas is exhausted into the atmosphere. Gasification of coal with air/ oxygen/ steam is an attractive option to avoid the ill-effects of fly ash in the combustion products. In fact, underground gasification of coal (UGC) involves the *in situ* conversion of coal into combustible gaseous products containing a mixture of carbon monoxide, carbon dioxide, hydrogen, methane and nitrogen.

During UCG, the coal in the seam reacts at elevated temperatures and pressures to produce gaseous combustible products (Figure 1). The gas composition, as well as the heating value depends on the thermodynamic conditions, grade of coal, and the gasifying agent composition and temperature. The gasification process starts with the ignition of coal seam at the beginning of the gasification channel. After ignition, the gasification agents move the process forward, along the gasification channel towards the production well. The main gasification reactions occur both in the solid and gaseous phases as well on their boundaries. The solid phase undergoes a drying process first to release the water evaporated from the coal; at a slightly higher temperature, pyrolysis processes give rise to some volatile gases. Following the pyrolysis, some homogeneous and heterogeneous reactions occur whose rates primarily depend on temperature (Figure 2). The homogeneous reactions occur completely in the gas phase, while the heterogeneous reactions occur at the solid (coal) surface.

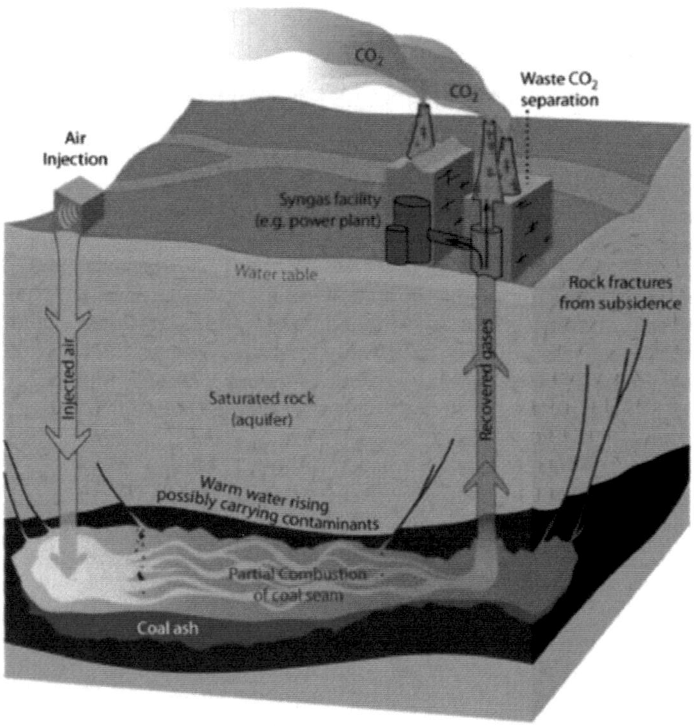

Figure 1. Principle of the UCG process [1]

2. Modelling of coal gasification

2.1. Heterogeneous reactions

Char (solid carbon) is consumed in the following heterogeneous reactions [2-4]:

differences in composition for some Indian coal samples vis-à-vis non-Indian coals are listed in Table 1.

2.2. Heterogeneous reactions

The homogeneous reactions arise from the gas phase processes such as the burning of volatile gases (taken in a general chemical form of CH_xO_y) and the reactions between the gaseous products such as CO, CO_2, H_2O and H_2. Some of the most important homogeneous reactions are listed below:

$$CH_xO_y + a\,O_2 \rightarrow CO_2 + b\,H_2O \quad (2.a)$$
$$CH_xO_y + c\,O_2 \rightarrow CO + d\,H_2O \quad (2.b)$$
$$CO + ½\,O_2 \leftrightarrow CO_2 \quad (2.c)$$
$$H_2 + ½\,O_2 \leftrightarrow H_2O \quad (2.d)$$
$$CO + H_2O \leftrightarrow CO_2 + H_2 \quad (2.e)$$

Where $a = 1 + x/4 - y/2$, $b = d = x/2$ and $c = (1 - y)/2$, based on stoichiometric balance. The kinetic parameters for the gas-phase reactions have been reported [5-7]. For reversible reactions, equilibrium data (Table 2) is often used to obtain the rates of the backward kinetic steps.

Table 2. Chemical equilibrium constants for carbon reactions [8, 9]

T, K	**1/T, K^{-1}**	**ln K$_p$**					
		1.a	$C+O_2=CO_2$	1.c	1.b	2.e	1.d
300	0.003333	23.93	68.67	-15.86	-20.81	4.95	8.82
400	0.0025	19.13	51.54	-10.11	-13.28	3.17	5.49
500	0.002	16.26	41.26	-6.63	-8.74	2.11	3.43
600	0.001667	14.34	34.40	-4.29	-5.72	1.43	2.00
700	0.001429	12.96	29.50	-2.62	-3.58	0.96	0.95
800	0.00125	11.93	25.83	-1.36	-1.97	0.61	0.15
900	0.001111	11.13	22.97	-0.37	-0.71	0.34	-0.49
1000	0.001	10.48	20.68	0.42	0.28	0.14	-1.01
1100	0.000909	9.94	18.80	1.06	1.08	-0.02	-1.43
1200	0.000833	9.50	17.24	1.60	1.76	-0.16	-1.79
1300	0.000769	9.12	15.92	2.06	2.32	-0.26	-2.1
1400	0.000714	8.79	14.78	2.44	2.80	-0.36	-2.36

In a typical UCG operation, the heterogeneous reaction of Eq. (1.a) and the homogeneous reactions of (3.a), (3.b), (3.c) and (3.d) may occur at the beginning of the gasification channel, forming what is known as the 'Oxidation Zone'. The endothermic Boudouard's reaction involving CO_2 reduction with char (Eq. 1.b), steam decomposition reaction (Eq. 1.c), the exothermic coal hydration or 'methanation' reaction (Eq. 1.d) may occur towards the later portion of the gasification channel, forming what is known as the 'Reduction Zone'. Over the whole length of the gasification channel, an important gas phase reaction such as the water-gas reaction (Eq. 3.e) can occur. Thus, UGC needs to be considered as a combination of complex flow, heat and mass transfer processes involving several gas phase and solid phase reactions. Often, simplified studies conducted on ground can provide the necessary data for understanding the complex phenomena in UGC.

In the Mechanical Engineering Department of IIT Madras, research on surface gasification of coal has been carried out in the last 10 years. A 20 kW laboratory scale entrained flow gasifier and a 100 kW bubbling bed gasifier have been set up, for gasifying high ash Indian coals (see Figures 3 and 4). The characteristics of Indian coal gasification have been compared with those of Australian and Japanese coals of low ash content. CFD simulations have also been performed to analyze the gasification phenomena, using the models presented above and the results have been published as research papers in archival journals [10-13]. The results obtained in our surface gasification studies of Indian coal are summarized in the next section.

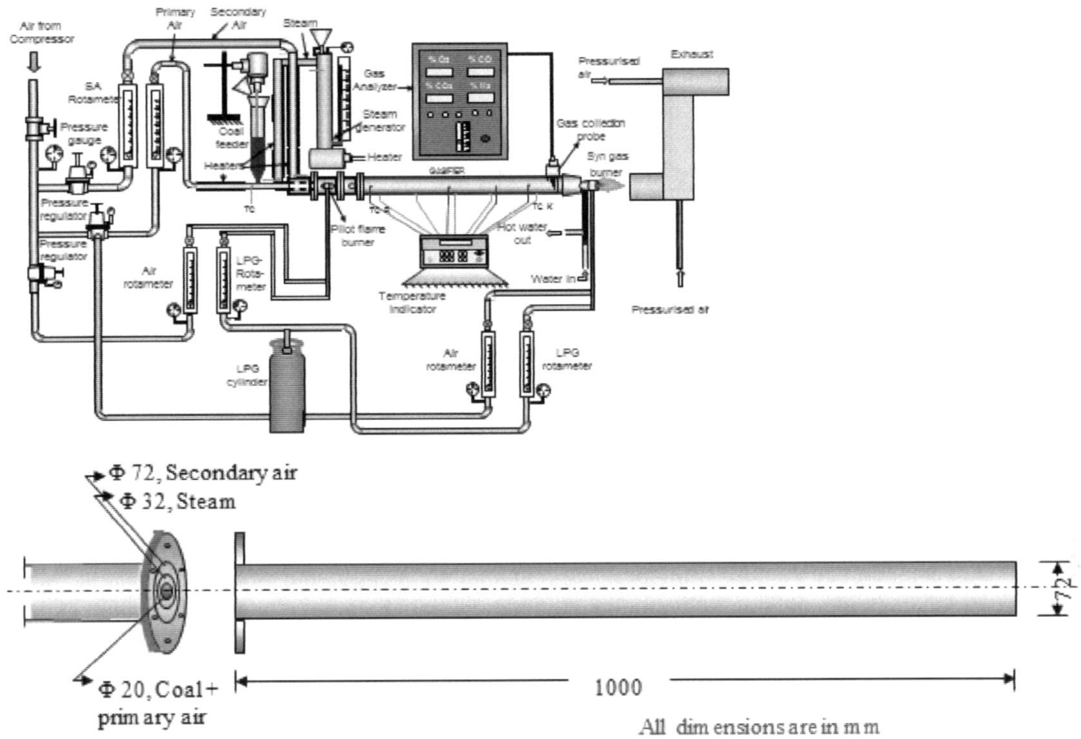

(a)

(b)

Figure 3. (a) Laboratory scale 20 kW entrained flow coal gasifier experimental set-up and (b) the corresponding computational domain [10] (tube of 1000 mm length)

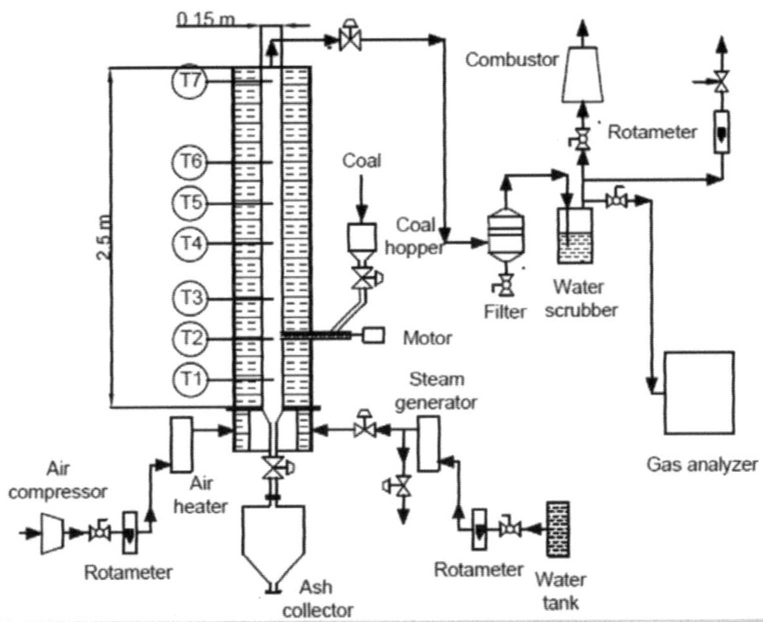

Figure 4. Pilot scale (~100 kW) bubbling bed gasifier and thermocouple locations [11]

Sl. No.	T1	T2	T3	T4	T5	T6	T7
Thermocouple Location (m)	0.2	0.5	0.8	1.2	1.5	1.8	2.8

3. Results and discussion

The laboratory scale entrained gasifier was operated with three grades of Indian coal and its gasification characteristics were compared with those of a typical Australian coal. The compositions of these coals are provided in Table 1. As a first exercise, the predictions of the theoretical model have been compared with available experimental results for an Australian coal and a Japanese coal, in Figure 5 and the associated Table 3. Very good match is observed between the numerical and experimental results in this validation exercise.

Table 3. Comparison of present numerical results with experimental results [10]

Gasification index	Australian coal		Japanese coal	
	Numerical *AR= 0.358*	*Experimental* *AR= 0.358*	*Numerical* *AR=0.392*	*Experimental* *AR=0.392*
CO	0.226	0.220	0.176	0.182
H_2	0.073	0.078	0.075	0.074
CO_2	0.054	0.060	0.077	0.065
CC (%)	71.88	66.81	75.02	72.46
HHV (MJ/Nm^3)	3.832	3.890	3.352	3.887
CGE (%)	51.33	52.50	49.63	53.23

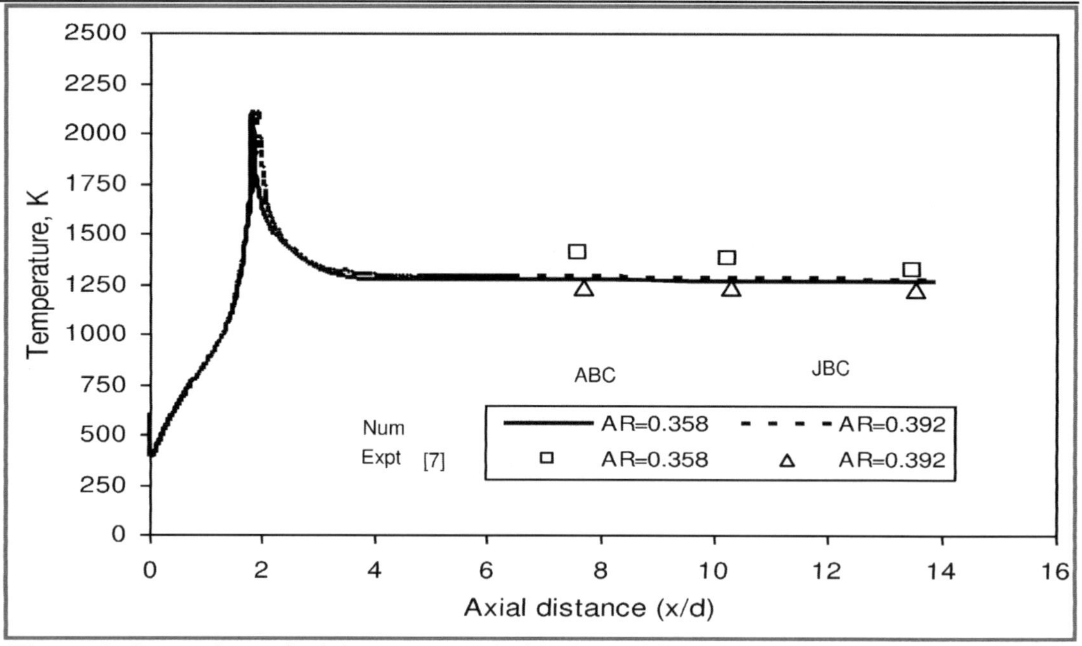

Figure 5. Comparison of axial temperature inside the gasifier with experimental results [10]

The effect of pressure on the evolution of gasifier temperature profile is depicted in Figure 6. It is evident that at higher pressure, the oxidation reactions occur at an earlier axial distance, due to the larger concentration of oxygen (per unit volume) in the mixture. The corresponding peak temperatures are also slightly higher. However, the evolution of volatiles is delayed at a higher pressure, as seen in Figure 7. Also, volatile evolution is relatively slower for a larger coal particle than for a relatively smaller particle, evidently due to the diffusion resistance offered by the ash layer. In Figures 8 and 9, it is illustrated that the formation of CO and H_2 are also accelerated at higher pressures. Therefore, gasification parameters such as carbon conversion (CC), higher heating value of the gas (HHV) and the cold gas efficiency (CGE) register higher magnitudes at higher system pressures as shown in Table 4. The steam ratio has only a minor effect on all gasification parameters for the high ash Indian coal, as seen from Table 5. Figure 10 shows that the numerically predicted product gas composition agrees fairly well with the measured gas composition at different air ratios considered.

In Table 6, the conditions studied in the 100 kW pilot scale gasifier are listed. The transient operation of the pilot scale gasifier and the corresponding exhaust temperature variations are shown in Fig. 11. The corresponding gas composition variation with time after an initial transient period of 2 hours is shown in Fig. 12. The average exhaust gas composition for various coal batch sizes is shown in Fig. 13. These studies illustrate that CC is more that 97% in the pilot scale combustor.

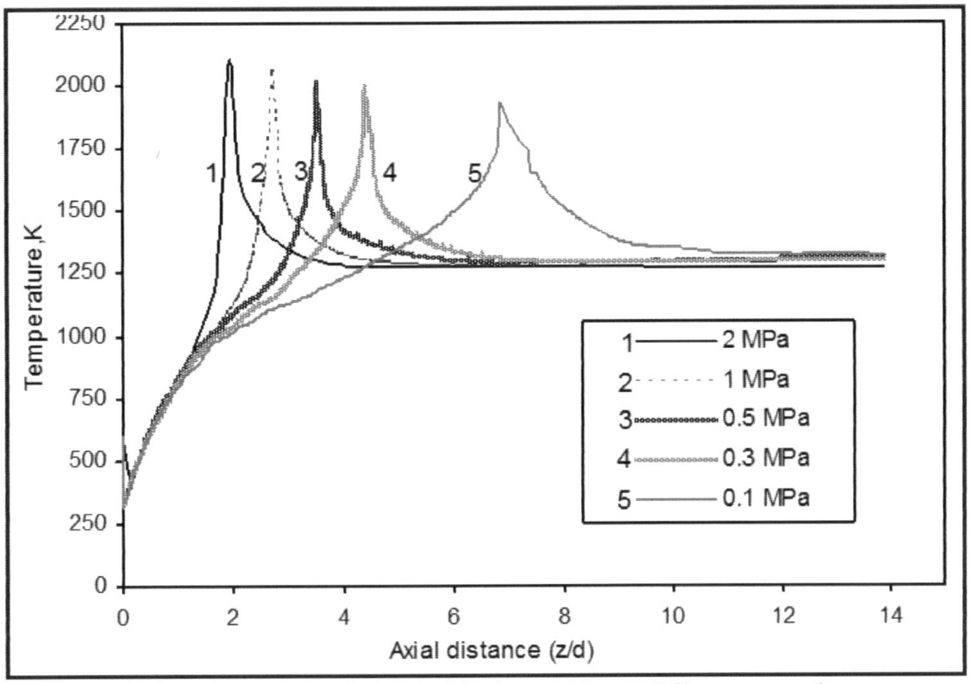

Figure 6. Axial temperature variation inside the gasifier at different operating pressures [10]

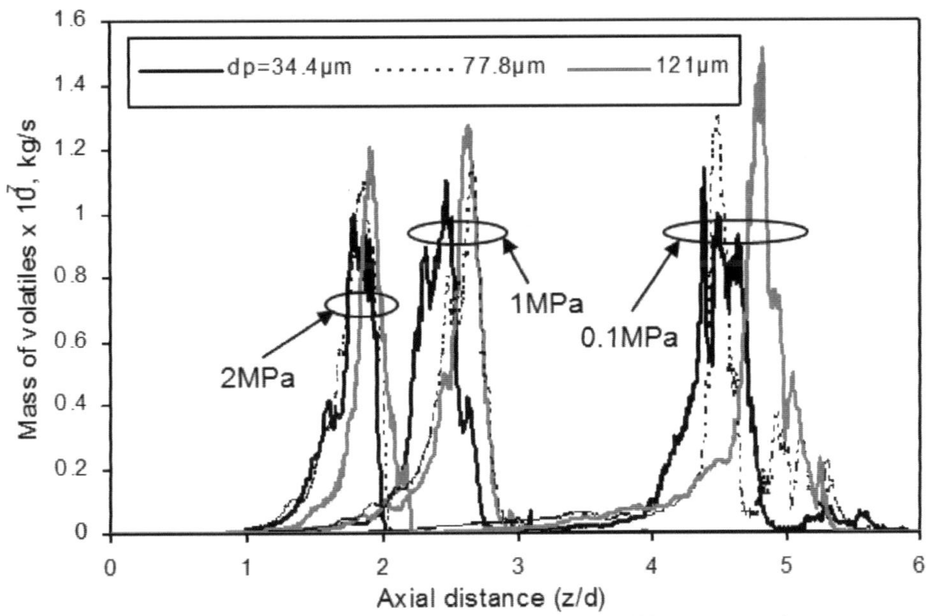

Figure 7. Devolatilization of representative particles at different operating pressures [10]

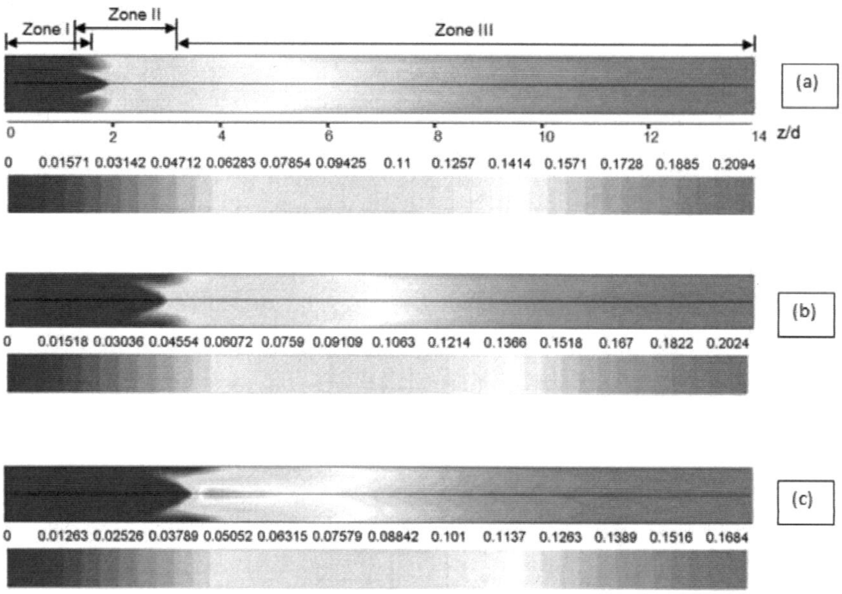

Figure 8. Evolution of CO with axial length in entrained gasifier (a:2 MPa, b:1 MPa, c:0.5 MPa) [12]

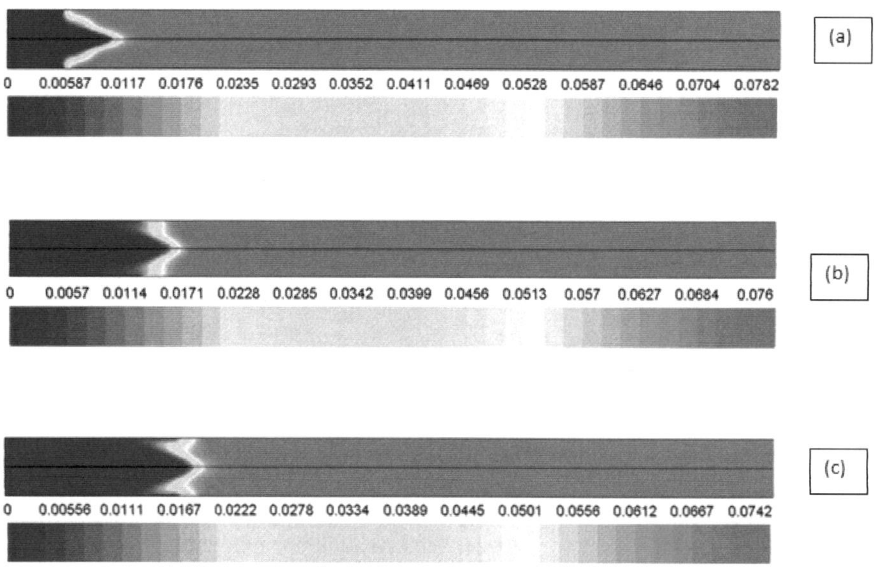

Figure 9. Evolution of H_2 with axial length in entrained gasifier (a:2 MPa, b:1 MPa, c:0.5 MPa) [12]

Table 4. Effect of pressure on gasification parameters for Indian coal- A (SR = steam ratio) [10]

Pressure (MPa)	**Carbon conversion CC (%)**		**HHV (MJ/m^3)**		**Cold Gas Efficiency (%)**	
	SR=0	*SR=0.1*	*SR=0*	*SR=0.1*	*SR=0*	*SR=0.1*
2	71.12	69.89	3.543	3.579	47.58	48.03
1	69.32	68.17	3.415	3.373	45.71	46.34
0.5	65.51	64.88	2.912	3.079	41.47	43.04
0.3	62.27	60.86	2.691	2.728	38.55	39.98
0.1	57.85	56.06	2.189	2.235	33.46	34.24

Table 5. Effect of steam ratio on gasification of Austalian and Indian coals, after [10]

Steam Ratio (SR)	CC (%)		HHV (MJ/Nm^3)		CGE (%)	
	ABC	*IC-A*	*ABC*	*IC-A*	*ABC*	*IC-A*
0.0	74.02	71.12	3.641	3.543	52.26	47.58
0.1	73.15	69.89	3.732	3.579	51.02	48.03
0.2	72.62	67.11	3.668	3.552	50.11	47.22
0.3	71.06	66.02	3.602	3.504	49.35	45.41

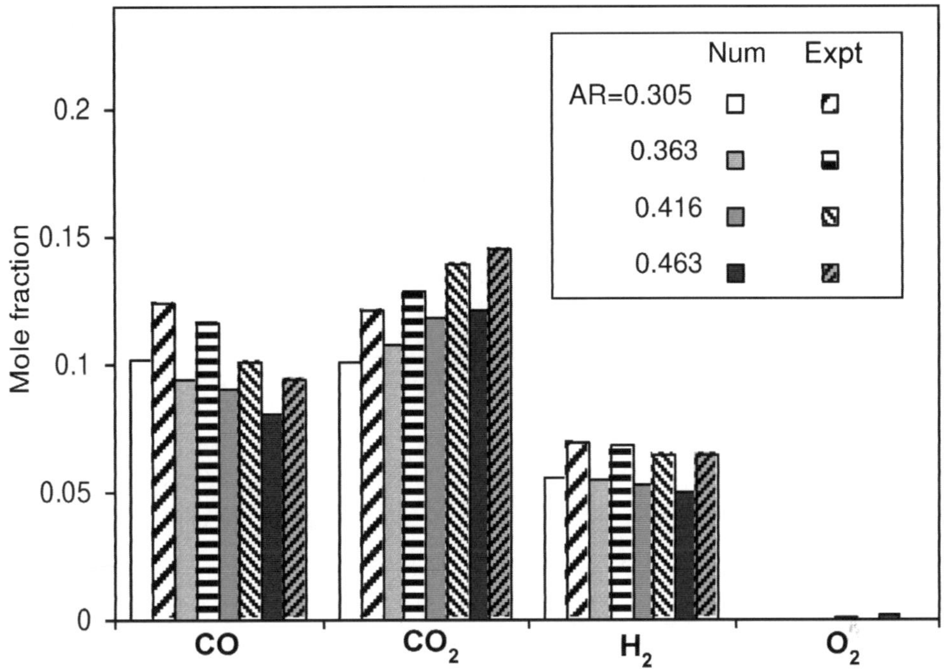

Figure 10. Comparison of predicted and measured product gas composition [12]

Table 6. Experiments conducted on the pilot scale bubbling bed gasifier [11]

Test run	Power output	Coal feed rate	Air flow rate	ER	Steam flow rate	Steam/coal ratio
	kW	*kg/h*	*kg/h*		*kg/h*	
1	100	20.24	40.6	0.35	-	-
2	80	16.19	32.48	0.35	-	-
3	80	16.19	27.84	0.30	-	-
4	80	16.19	23.20	0.25	2.42	0.15
5	80	16.19	23.20	0.25	3.23	0.25
6	80	16.19	23.20	0.25	4.05	0.30

4. Conclusion

The present experimental and modeling studies illustrate that gasification processes of high ash Indian coals can be modeled fairly accurately. Operation of the gasifier at higher pressure is definitely advantageous to achieve more efficient gasification. As compared to the foreign coals with lower ash content, Indian coals give relatively poorer efficiency during gasification. However, the advantage of not taking the ash out of the mine is greatest for the Indian high ash coals. Although modeling of surface gasification has been presented in this paper, the methodology outlined here can be combined with the flow, heat transfer and mass transfer processes occurring in UGC, for the development of comprehensive gasification models.

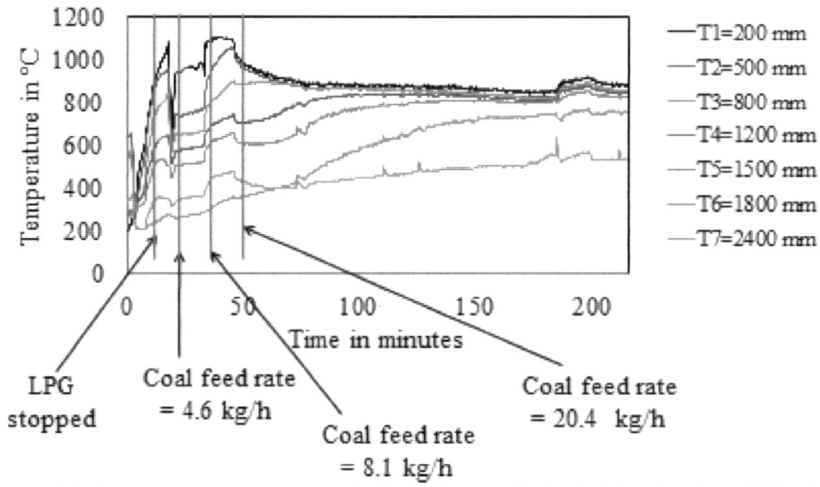

Figure 11. Transient operation of the pilot scale bubbling bed gasifier [11]

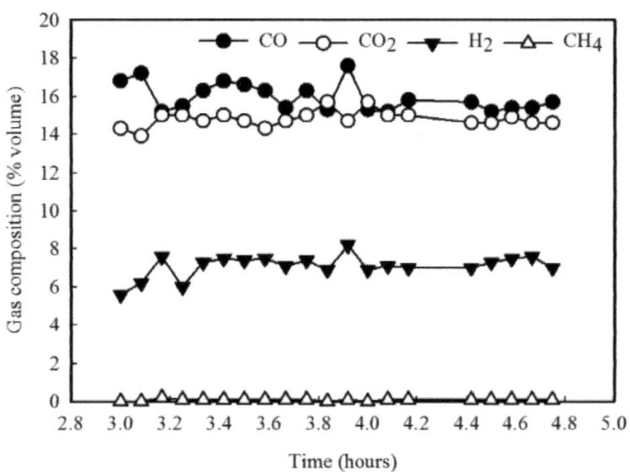

Figure 12. Gas composition variation with time for pilot scale gasifier [11]

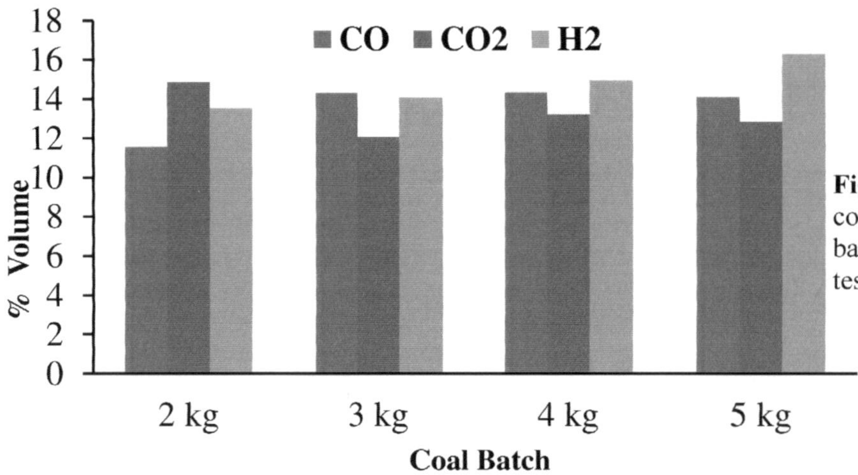

Figure 13. Average gas composition for different batch coal gasification tests

References

[1] Laciak M, Kostúr K, Durdán M, Kačur J and Flegner P 2016 *Energy* **114** 332-343
[2] Smith I W 1982 *Nineteenth Symposium (International) on Combustion* (The Combustion Institute) p 1045-1065
[3] Chen C, Horio M and Kojima T 2000 *Chemical Engineering Science* **55** 3861-3874
[4] Chen C, Horio M and Kojima T 2000 *Chemical Engineering Science* **55** 3875-3883
[5] Shaw D W, Zhu X, Misra M K and Essenhigh RH 1991 *23rd Symposium (International) on Combustion* (The Combustion Institute) p 1155-1162
[6] Westbrook C K and Dryer F L 1981 *Combustion Science and Technology* **27** 31-43
[7] Watanabe H and Otaka M 2006 *Fuel* **85** 1935-1943
[8] Lee S and Shah Y T 2012 *Biofuels and Bioenergy: Processes and* Technologies (Boca Raton: CRC Press)
[9] Walker P L, Rusinko F and Austin L G 1959 *Advances in catalysis* **11**133-221
[10] Ajilkumar A, Shet U S P and Sundararajan T 2010 *Heat Transfer Engineering* **31** 495-508
[11] Kumar K V, Bharath M, Raghavan V, Prasad, B V S S S, Chakravarthy S R and Sundararajan T 2017 *Applied Thermal Engineering* **116** 372-381
[12] Ajilkumar A, Sundararajan T and Shet U S P 2009 *International Journal of Thermal Sciences* **48** 308-321
[13] Singh N, Raghavan V and Sundararajan T 2014 *International Journal of Energy Research* **38** 737-754

IOP Conf. Series: Earth and Environmental Science **76** (2017) 012007 doi:10.1088/1755-1315/76/1/012007

Performance, cost and environmental assessment of gasification-based electricity in India: A preliminary analysis

Abha Rani[1,]*, Udayan Singh[2], Jayant[3], Ajay K Singh[4], Siba Sankar Mahapatra[2]

[1] Department of Applied Mathematics, Indian Institute of Technology (Indian School of Mines), Dhanbad, INDIA
[2] Department of Mechanical Engineering, National Institute of Technology Rourkela, Rourkela, INDIA
[3] Department of Chemical Engineering, Manipal Institute of Technology, Manipal, INDIA
[4] CSIR-Central Institute of Mining and Fuel Research, Dhanbad, INDIA

* Corresponding Author. Email: ajayabha@yahoo.com

Abstract. Coal gasification processes are crucial to decarbonisation in the power sector. While underground coal gasification (UCG) and integrated gasification combined cycle (IGCC) are different in terms of the site of gasification, they have considerable similarities in terms of the types of gasifiers used. Of course, UCG offers some additional advantages such as reduction of the fugitive methane emissions accompanying the coal mining process. Nevertheless, simulation of IGCC plants involving surface coal gasification is likely to give reasonable indication of the 3E (efficiency, economics and emissions) prospects of the gasification pathway towards electricity. This paper will aim at Estimating 3E impacts (efficiency, environment, economics) of gasification processes using simulation carried out in the Integrated Environmental Control Model (IECM) software framework. Key plant level controls which will be studied in this paper will be based on Indian financial regulations and operating costs which are specific to the country. Also, impacts of CO_2 capture and storage (CCS) in these plants will be studied. The various parameters that can be studied are plant load factor, impact of coal quality and price, type of CO_2 capture process, capital costs etc. It is hoped that relevant insights into electricity generation from gasification may be obtained with this paper.

1. Introduction

Coal is the mainstay of the Indian power sector. While the net installed capacity of India's power sector is more than 315 GW, around 60 percent is from the coal-fired power plants [1]. The dominance of coal has prevailed since independence and has strengthened over various five year plan (FYP) periods. The installed capacity within the coal sector has seen a seventy times increase between the end of the 11th FYP and the 1st FYP [2]. This surge has resulted in large-scale electrification for the cities and villages of the country. Coal production has also been enhanced and it is envisaged to increase the production by Coal India Ltd to 1 billion tonnes by 2019-20 [3]. While increases in coal production as well as power generation are planned and projected, sustainability in the power sector also needs to be targeted.

It should be noted that the power sector is the key contributor to greenhouse gas (GHG) emissions from India. India's total CO_2 emissions (without LULUCF) in 2010 were about 1575 Tg, out of which 815 Tg were from power sector itself [4]. It has also been reported that significant SO_2, NO_x and PM pollution also results from power plants in India. The power sector thus has an important role to play in the reduction of gaseous and particulate local air pollutants. A number of key initiatives are being undertaken in this regard with active support of the Government of India. These include retirement of old and inefficient units, renovation and modernization (R&M), installation of supercritical units, and so on.

Globally, a key segment of clean coal initiatives is gasification of coal. Gasification leads to a reduction of emissions and the power generation is considered to be superior to that from pulverized coal power plants. In this paper, a highlight of various Indian works based on coal gasification shall be presented. A discussion on the current and future trajectory of use for coal and natural gas in India has also been undertaken.

1.1. Objectives and scope of this paper

This paper begins with a review of the prior work done on plant-level analysis for coal gasification, specifically in the Indian context. A background of the role of India's emerging energy options vis-à-vis coal and natural gas are also covered. Subsequently, we adopt a simulation based approach to estimate the cost of electricity, atmospheric emission and net plant efficiency for plants employing coal gasification. The performance of such plants is compared to the plants employing pulverized coal-firing plants. The impact of CO_2 capture on gasification based plants has been studied in details. A parametric analysis of the various factors involved to aim at cost reduction has been performed.

It should be noted here that this is a simulation-based analysis focusing only upon the power plant, and therefore no discussion has been made regarding the mining of coal or its in-situ gasification. The paper is therefore just an indicative analysis for the feasibility of coal gasification and significant improvements may be made to the analysis. Further, the analysis has been done only for north-eastern (tertiary coal).

2. India's coal sector: an overview

A large part of India's primary energy consumption as well as power requirements is contributed to by the combustion of coal. Coal and lignite reserves in the country stand at a total of more than 350 billion tons. The annual production of coal in 2015-16 was about 640 million tons, but recently the Indian coal giant, Coal India Limited envisaged production of 1 billion tones by the end of the decade. Coal is responsible more meeting of more than 70 percent of primary energy demands [5].

On the supply side, coal has seen a significant increase in terms of the net installed capacity of coal-fired power plants. From an installed capacity of ~750 MW at the installed capacity at the time of independence, an increase of more than 250 times has been seen [2]. The growth has resulted in considerable success in terms of India's developmental objectives. With the indication of the power ministry indicating that India is now a power-surplus nation, a transition towards a developed country will require affordable, adequate and sustainable electricity for all. Most modelling studies suggest that this will be accompanied by a further growth of coal-based electricity generating units (EGUs) [6, 7]. Shukla et al [8] project that in a business-as-usual (BAU) scenario; coal will remain the mainstay of India's electricity sector upto 2100. They also project that coal will dominate the solid fuels mix, but its share may decrease over time due to climatic targets.

Multiple modelling studies have during the past decade forecasted the trend of coal production in the future. Patzek and Croft [9] have suggested that coal production is nearing its peak globally (also in India) and will reach its peak production in 2011, in terms of energy production. On the basis of multi-Hubbert cycle analysis, they have suggested that the annual peak of CO_2 emissions from coal use shall be 1.2 Gt/year. Mohr and Evans [10] have forecasted that coal production upto 2100. They project that bituminous coal production will peak in 2046 (at about 900 Mt/year). Similarly, lignite production has been projected to peak in the year 2010 at about 40 Mt/year. An interesting point to note here is that while the bituminous peak coal production has been predicted to be less than 900 Mt annually, CIL already intends to produce more than a billion tons of coal by 2020. Thus, the

IOP Conf. Series: Earth and Environmental Science **76** (2017) 012007 doi:10.1088/1755-1315/76/1/012007

projections for coal production are highly dynamic and largely dependent upon energy demands in the country.

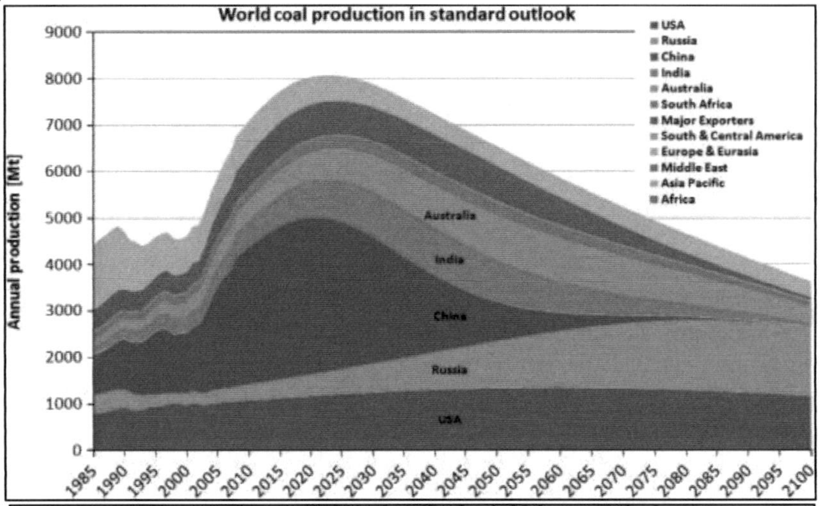

Figure 1. Standard case outlook for future global coal production. Source: Ref. [11]

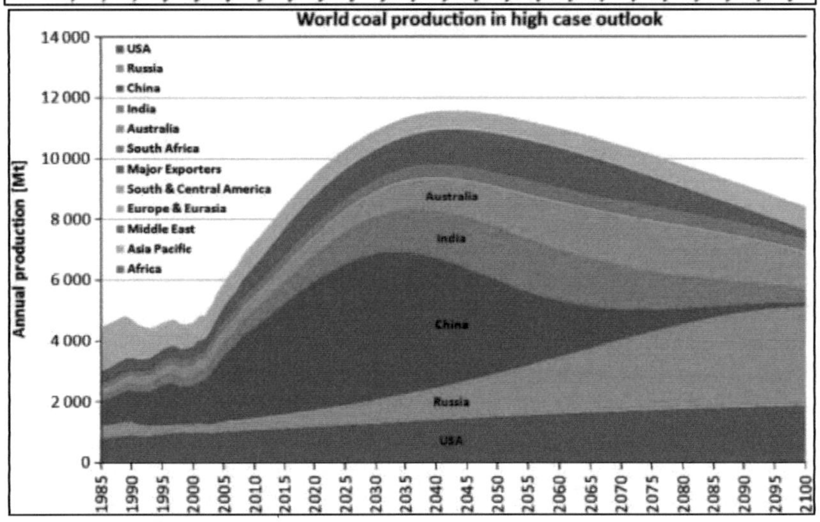

Figure 2. Even if the world's coal reserves are twice as large as reported this does not postpone the peak in production more than two decades. Source: Ref. [11]

In another similar work, Höök et al [11] have projected that the current leading producers of coal shall retain the driving seat in the future as well. Figures 1 and 2 show global coal production in the upcoming years in the standard and high-production outlooks. The projection suggests that while Chinese coal production will drastically reduce near 2050-2060, Indian and Australian coal production will remain more or less steady at least upto 2070. Thus, it can be seen that different researchers have given widely varying estimates of coal production and peaking for India. A key facet of how long coal can be continued to be used is very much dependent upon the degree of sustainability built-in with the technology of coal usage. The unconventional gas based industry (mostly, coalbed methane (CBM) and underground coal gasification (UCG)) will play a key role in the decarbonization of coal industry, which will permit long-term use of coal even in midst of stringent climatic targets.

Similarly, the implementation of CO_2 capture and storage (CCS) technologies will play a key role in ensuring coal-based energy security for India in the long term [6]. Researchers have given their estimates on the role CCS may play in the decarbonization of the Indian power sector. Shukla et al [12] suggest that there may be two possible means towards reduction of overall power sector carbon intensity. One may be through an increased carbon tax (conventional pathway) having significant reliance upon CO_2 sequestration, while the other (sustainable pathway) will focus mostly on the increased deployment of renewable power. Singh et al [13] attempted to quantify the monetary benefits associated with CCS in a high carbon price scenario. They projected in an CCS-intensive scenario with high carbon tax, the monetary benefits of CCS would be of the order of US$ 1.2-1.5 trillion to avoid 11,000 Mt-CO_2 over the years 2030-2050. Nevertheless, significant challenges remain for installation of CO_2 capture units in India. Singh and Rao [14] have estimated a cost of avoidance of US$ 60-85/t-CO_2 if the technology is to be utilized. Also, as Singh et al [15] have pointed out, some societal perspectives see CCS as a technology which does more harm than good. Therefore, a dedicated effort needs to be put in to reduce energy penalty and costs with the technology before it reaches anywhere close to commercialization. One possible way can be with gasification of tertiary northeastern coals, and also of lignite in Rajasthan and Gujarat. This route will be studied in this paper.

3. Coal gasification: previous studies

Initial efforts towards coal gasification started in the erstwhile Central Fuel Research Institute [16]. Krishnudu et al [17, 18] performed studies on a moving bed gasifier for Indian coals. They delineated three zones (fast devolatilization, slow devolatilization with gasification and gasification) in the gasifier for studies based on low-rank coal. They also concluded that one-third of the bed height was utilized for gasification and the rest for devolatilization. Similarly, in a parametric study to determine the effect of pressure variation on three south Indian coals. They estimated the gas yield of the order of 1.4-1.7 m_n^3/kg-coal. Iyengar and Haque [19] demonstrated the loss of efficiency with increase in ash content. Chatterjee et al [20] also carried out simulation and experimental work on laboratory scale and showed a significant degree of correlation between the same.

With developed numerical investigation techniques, many researchers have tried to simulate coal gasification processes. Ajilkumar et al [21] modelled entrained lab-scale gasifier using Fluent software. They predicted carbon conversion, gas heating value and cold gas efficiency with reasonable accuracy. Ajilkumar et al [22] also simulated tubular gasifier and observed strong decrease in performance with rise in ash content. Singh et al [23] tried to understand the effect of coal properties on gasification. They found that high carbon content and low ash content were instrumental for good performance of gasifier. Singh et al [24] carried out further analysis for optimum performance of the gasifier. They suggested that steam-to-coal ratio of 0.4 and 1.5 for optimal for air-based and oxygen-based systems.

Datta et al [25] attempted to study the agglomeration behaviour of Indian coals. They found that agglomeration was found at temperature close to 950^{0}C and ash fusion temperature was above 1200^{0}C. They concluded that compositional variations and Fe^{2+} presence led to the agglomeration. Chavan et al [26] developed multivariate regression (MVR) and artificial neural network (ANN) models to predict outputs from fluidized bed gasifiers under steady state condition. Patil-Shinde used ANN along with genetic programming to predict the amount and quality of syngas produced [27].

Some studies have also been carried out for underground coal gasification in India. Laboratory studies on cavity growth were performed by Daggupati et al [28]. They also attempted to study product gas composition in Indian context [29]. Mandapati et al [30] further approached the problem with kinetic modelling techniques and developed a one-dimensional reaction diffusion model. Bhaskaran et al [31] further compared gasification characteristics of lignite and hard coal and found that the former was ten times more reactive and also yielded syngas with ~2.5 times more calorific value than the latter. Sanga et al [32] developed a model using COMSOL Multiphysics for chemical reactions, through characterization of coal from Kapurdi mine in Rajasthan.

3.1. Gasification with CCS

CO_2 capture using pre-combustion techniques can be performed in gasification based plants. Such plants can led to affordable electricity at about € 77/MWh, even when operated with CO_2 capture [33]. This cost effectiveness of gasification plants (both UCG-CCS and IGCC-CCS) can be of use to the Indian energy sector. Several studies have been performed for these technologies to assess them techno-economically [33-38]. Table 1 shows some of these studies.

Table 1. A review of some gasification studies (both surface and underground coal gasification) performed globally – with inclusion of CO_2 capture and storage

Reference	Inputs / Assumptions/ Methodology	Results/ Conclusions
Olateju and Kumar [34] ***Hydrogen production using UCG with and without CCS***	• Plant model developed • Entrained flow gasifier with pressure = 120 bar • Injection temperature = 1227 deg. C • Total capital cost = CAD$ 1569 million • CO_2 flow rate = ~11000 t/day • Hydrogen capacity = 660,00 kg/day • Load factor = 85%	• Natural gas price is key determinant for competition between SMR and UCG for Hydrogen production • UCG shows specific cost advantage when CCS is implemented • Internal rate of return for success of UCG is less than 4.6-5.4 • H_2 production cost \$ 1.78/kg-H_2 without CCS and \$ 2.1-2.7/kg-H_2 with CCS • UCG is most cost-effective at gas prices more than \$ ~5.3/GJ • Risk of investment lower for UCG due to minimal fuel cost
Eftekhari et al [35] ***Exergy analysis of UCG with CCS***	• A chemical equilibrium model developed • The system is considered to be at a constant (user defined) pressure. • Intitial temperature= 35deg. C • Adiabatic conditions are assumed • CaO (particle size= 50μm) injected into the UCG cavity to favour H_2 production • Effect of oxygen/water on product composition is investigated • Wollastonite is reacted with CO_2 at 80 bar and 500K to produce H_2 and CO	• Conventional UCG with oxygen gives a higher recovery factor than air gasification. • Zero emission recovery factors of deep UCG are negative • Use of CaO in UCG to reduce CO_2 emission is not feasible. • Calcium silicates showed better recovery than CaO • Optimum practical recovery factor of 67% is achieved with CO_2 emission of 0.174kg/MJ • Whereas, a minimum CO_2 emission value of 0.022Kg/MJ can achieved by keeping recovery factor at 47%
Sheng et al [33] ***Feasibility study of deep UCG in Bulgaria***	• UCG feasibility study conducted on deep coal seams • Coal seams selected for study were 1000-1500m deep that contained 75MT of UGC compliant coal in them • Finite element software package ABAQUS was to model the thermo-geomechanical behaviour • The elastic modulus of the coal is gradually decreased with temperature to simulate the mechanical failure of coal after gasification • 3D thermo-mechanical model incorporating real lithological structures and other properties was	• The safety distance of at least 150m is proposed between the gasification channels and geologic faults • UCG-CCS operations were shown to have a little effect on the ground water quality. • UCG injections wells instead of production wells to be used for CO_2 injection to prevent corrosion of the production well • UCG-CCS concluded to be economic

Reference	Inputs / Assumptions/ Methodology	Results/ Conclusions
	developed. • The model was further used to study the faults, the stresses and the permeability around the UCG cavity. • A hydrogeological model was also prepared, for the 344 wells where examined for spreading data and hydrogeological parameters.	
Cormos [36] ***IGCC power generation with CCS***	• Two entrained flow gasifiers (Shell and Siemens) were studied with and without CCS • M701G2 gas turbine used to produce 400-500 MW • CCS case studies are designed to capture more than 90% of the carbon • Modelling and Simulation are done using ChemCAD and Thermoflex software packages. • Developed model was validated against industrial data	• Introduction of CCS costed 7-9% energy penalty • Shell type gasifier shows better overall net electrical efficiency • Siemens type gasifier showed better hydrogen efficiency • Though the siemens type gasifier shows greater net efficiency • Shell type gasifier required lesser overall capital and operation and maintenance cost.
Siefert and Litster [37] ***Analysis of IGCC power plant***	• HSC Chemistry 6.0 software package was used to model GE entrained gasifier. • Slurry of crushed coal and water is prepared and pressurised to 4.2 MPa. • Gasifier exit temperature is 1200 deg. C • Water to carbon and oxygen to carbon ratio are kept at 0.46:1 • 0.5% H_2S (molar basis) is assumed in the economic model • Syngas undergoes WGS reaction at 250 deg. C and 4.5 MPa	• System efficiency of 43 % for IGCC-CCS-EOR is reported • IRR= 8 ± 4%/yr (at 50/MWh) • NGCC is reported to yield the maximum rate of return i.e. 22%/yr • Study finds it possible for advanced coal based power plants to achieve IRR= 8%/yr at load electricity price of $50/MWh
Kunze and Spliethoff [38] ***IGCC plant model for 2020***	• Aspen Plus and Ebsilon Professional software packages were used for modelling purposes • Coal and biomass are characterised in a yield rector • Redlich-Kwong-Soave method is selected as selected as global method in Aspen Plus • Gibbs reactor model is used for the model of reactive section • Radfrac columns are used to model vap-liq operation	• Syngas yield of 2200 and 1644 (STP)/Kg coal as obtained • 0.273kWh/kg O_2 of power consumption was seen for a pressure of 6 bar in high pressure column and 1.5 bar in low pressure column • A net plant efficiency of 38.5% for hard coal and 41.9% for lignite is seen.

4. **Comparison between combustion and gasification plants**

In this paper, the Integrated Environmental Control Model (IECM) is used as a software tool to simulate the power plants. Since, we intend to understand only the gasification process; the underground coal gasification is treated akin to surface coal gasification. Thus, a comparative assessment is made for pulverized coal (PC) plants and integrated gasification combined cycle plants (IGCC) using some assumptions that describe coal costs and other assumptions fit to the Indian conditions. It should be noted that plant component costs are assumed as default IECM costs. However, replacements are made to other assumptions. For detailed methodology, the reader may refer to Berkenpas et al [39] or other works pertaining to the use of IECM in Indian conditions [40-42].

In this study, we wish to simulate a plant operating with northeastern coal. Chaulya et al [43] have performed detailed coal characterization for number of tertiary coal samples. Further, as Table 2 shows, these samples have considerable similarity to some American coals. Thus, we assume the characteristics of Appalachian coal basin since it eases syngas prediction within the model's framework. The coal price is accordingly adjusted as per the norms and notification of the Coal India Limited.

Table 2. Similarity between Tertiary coals and American Coals (Appalachian and Illinois#6). Source: IECM Model and Chaulya et al [43]

Coal samples	**Proximate analysis(% wt.)**		**Ultimate analysis(% wt.)**			**C.V.(Kcal/kg)**
	M	Ash	C	H	N	
Ledo	2.8-3	5.1-18.2	54.8-70.2	2.1-6.6	1-1.2	6500-7075
Baragolai	2.1-3.5	6.3-14.4	66.3-69.8	4.7-5.1	1.2-1.4	6685-7025
Tirap	2-3.2	3.9-11.1	67.4-69.9	4.8-4.9	1.2-1.3	6900-7640
Tikak	2.3-2.8	3.4-13.1	64.3-75.2	4.6-5.4	1.1-1.3	6400-7600
Tipong	2.3-2.9	4.3-9.2	60.9-74.5	5-5.4	1-1.2	6935-7630
Dilli-Jeypore	4.4-7.5	1.4-9.0	75.5-79.4	5.3-6.7	1-1.3	6335-7395
Appalachian	5.05	7.24	73.81	4.88	1.42	7370
Illinois#6	11.12	9.70	63.75	4.50	1.25	6486

The price of raw materials is taken from Singh and Rao [40] and is shown in Table 3, alongwith some other important assumptions.

Table 3. Some important assumptions for simulation of IGCC plant

Paramater	**Value**
Load Factor	90
Gasifier Make	GE 7EA
Debt:Equity Ratio	70:30
Cooling system	Wet Cooling Tower (in view of new regulations)
Slag management	Landfill
H_2S control	Selexol
Raw material prices[a]	
Activated Carbon ($/t)	1215
Ammonia ($/t)	376.7
Caustic ($/t)	424.6
Urea ($/t)	455.7

[a] From Singh and Rao [40]

For the PC plant, the plant based assumptions are taken from Singh and Rao [40] with the same coal characteristics and costs as in the above case.

Without CCS, pulverized coal and gasification plants perform similar in terms of efficiency, while cost is slightly higher for IGCC. Gaseous emissions are lower by an order of magnitude for IGCC, thereby showing greater importance for air pollution. Makeup water is also significantly lower for gasification-based plants. With CCS, however, IGCC shows distinct advantage with tertiary coal in terms of cost and energy penalty. Water use increase in minor in IGCC plant, as compared to PC. In a low-carbon world, gasification will show distinct advantages. It may also be noted that Sulfinol serves dual purposes (for CCS and H_2S removal). The detailed results are shown in Tables 4 and 5.

Table 4. Results without CO_2 capture and storage

	Unit	**PC**	**IGCC**
Efficiency	%net	37.27	36.20
COE	$/MWh	71.92	83.17
CO_2 Emissions	kg/kWh	0.8541	0.8342
SO_2 Emissions	kg/kWh	2.473e-3	2.889e-4
NO Emissions	kg/kWh	3.860e-4	4.899e-5
NO_2 Emissions	kg/kWh	3.115e-5	3.953e-6
Make up Water	kg/kWh	2.499	1.554

Table 5. Results with CO_2 capture and storage

	Unit	**PC***	**IGCC#**
Efficiency	%net	26.45	30.56
COE	$/MWh	122.9	118.0
CO_2 Emissions	kg/kWh	0.1206	9.398e-2
SO_2 Emissions	kg/kWh	Very Low (Due to Assumptions)	3.875e-5
NO Emissions	kg/kWh		5.606e-5
NO_2 Emissions	kg/kWh		5.606e-5
Make up Water	kg/kWh	4.377	1.729

* Amine-based capture
\# Sour-shift + Selexol

Further, it is seen that air Separation Unit and CO_2 capture unit Contributes to 21% cost each. While the gasifier contributes to 54% of the costs (see Figure 3). Thus, process improvements should target gasifiers and indigenous development of gasifiers is a must.

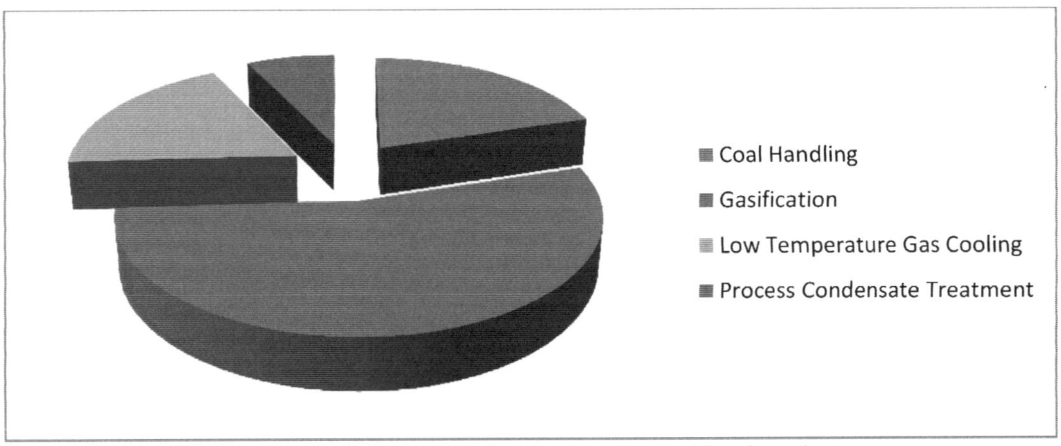

Figure 3. Capital cost distribution for gasification plant

The effect of increase in number of gas turbines also affects cost reduction, but is largely dependent on power output needed). Three turbines result in net 800 MW. An increase of gasification temperature by 56 deg. C (from 1343 to 1399), reduces COE by US$ 2/MWh. Joint Incentives in terms of carbon tax (w/CCS) or sulfur tax (w.o. CCS) can lead to higher inclination towards gasification technology. Gasification of tertiary coals can offer cheaper, cleaner means of electricity. If CCS is to be implemented, gasification of such coals can result in more feasibility (even at lower Carbon Price due to low CO_2 Emission Factor). Process development in gasification is the key to reducing costs and energy penalty for CCS.

Figure 4 shows the effect of increase in number of gas turbines. Increase in number of gas turbines leads to significant reduction in the cost of electricity. But it is also dependent upon the required output as three turbines result in net output of 800 MW. An increase of gasification temperature by 56 deg. C (from 1343 to 1399), reduces COE by US$ 2/MWh.

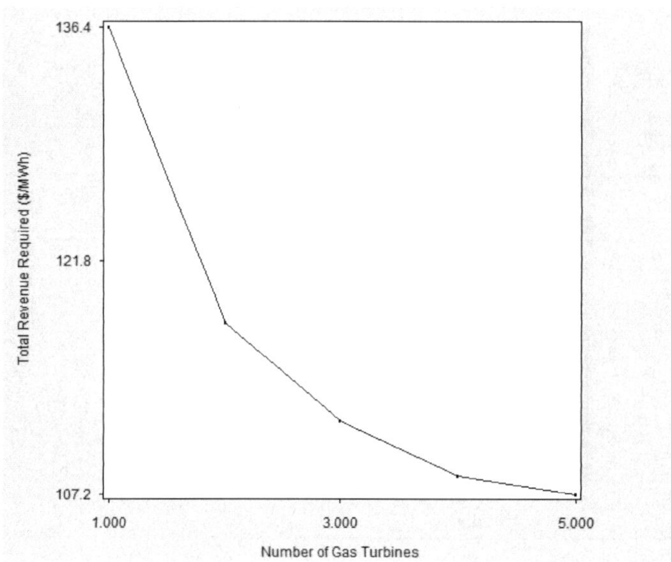

Figure 4. Relation of cost of electricity to number of gas turbines

5. Conclusion

Gasification of tertiary coals can offer cheaper, cleaner means of electricity. These coals have low-ash and high gross calorific value. If CCS is to be implemented, gasification of such coals can result in more feasibility (even at lower carbon price due to lower CO_2 emission factor). Further, significant environmental benefits in terms of air pollutants and water consumption can be seen in gasification based plants. It may be noted that process development in gasification is key to reducing costs and energy penalty for CCS. A potential question also arises – "Can avenues for gasification of lignites be opened up?". The work ahead should focus upon modelling of integrated UCG-CCS processes. Work on chemical looping based capture may also be taken up since it may offer large reduction in energy penalty.

References

[1] CEA 2017 *All India Installed Capacity – April 2017* available at <http://www.cea.nic.in/reports/monthly/installedcapacity/2017/installed_capacity-04.pdf>

[2] CEA 2015 *Growth of Electricity Sector in India From 1947-2015* available at <http://www.cea.nic.in/reports/others/planning/pdm/growth_2015.pdf>

[3] PIB 2015 *CIL Gearing Up For 1 Billion Tonne Coal Production Mark* available at <http://pib.nic.in/newsite/PrintRelease.aspx?relid=115635>

[4] MoEFCC 2016 *India – First Biennial Update Report to UNFCCC* (New Delhi: Ministry of

Environment, Forests and Climate Change, Government of India)
[5] CSO 2017 *India Energy Statistics* (New Delhi: Ministry of Statistics and Power Implementation)
[6] Garg A and Shukla P R 2009 *Energy* **34** 1032-1041
[7] Chikkatur A P, Sagar A D and Sankar T L 2009 *Energy* **34** 942-953
[8] Shukla P R, Dhar S, Pathak M, Mahadevia D and Garg A 2015 *Pathways to deep decarbonization in India* (SDSN-IDDRI)
[9] Patzek T W and Croft G D 2010 *Energy* **35** 3109-3122
[10] Mohr S H, and Evans G M 2009 *Fuel* **88** 2059-2067
[11] Höök M, Zittel W, Schindler J and Aleklett K 2010 *Fuel* **89** 3546-3558
[12] Shukla P R and Dhar S 2016 *Enabling Asia to Stabilise the Climate* (Springer Singapore) p 41-54
[13] Singh U, Rao A B and Chandel M K 2017 *Energy Procedia* DOI: 10.1016/j.egypro.2017.03.1896
[14] Singh U and Rao A B 2016 *Energy Procedia* **90** 326-335
[15] Singh U, Sharma N, Mahapatra S S 2016 *International Journal of Coal Science and Technology* **3** 215-225
[16] Abrol D 2011 *Science and Modern India: An Institutional History, c.1784-1947: Project of History of Science, Philosophy and Culture in Indian Civilization* Vol 15, Part 4, ed U Dasgupta (Pearson) p 185-231
[17] Krishnudu T, Madhusudhan B, Narayan Reddy S, Sastry V S R, Seshagiri Rao K and Vaidyeswaran R 1989 *Industrial and Engineering Chemistry Research* **28** 438-444
[18] Krishnudu T, Madhusudhan B, Narayan Reddy S, Seshagiri Rao K and Vaidyeswaran R 1989 *Fuel Processing Technology* **23** 233-256
[19] Iyengar R K, Haque R 1991 *Fuel Processing Technology* **27** 247-262
[20] Chatterjee P K, Datta A B and Kundu K M 1995 *The Canadian Journal of Chemical Engineering* **73** 204-210
[21] Ajilkumar A, Sundararajan T and Shet U S P 2009 *International Journal of Thermal Sciences* **48** 308-321
[22] Ajilkumar A, Sundararajan T and Shet U S P 2007 *ASME/JSME 2007 Thermal Engineering Heat Transfer Summer Conference collocated with the ASME 2007 InterPACK Conference* (Vancouver) p 297-305
[23] Singh N, Raghavan V and Sundararajan T 2013 *9th Asia-Pacific Conference on Combustion* (Gyeongju)
[24] Singh N, Raghavan V and Sundararajan T 2013 *International Journal of Energy Research* **38** 737-754
[25] Datta S, Sarkar P, Chavan P D, Saha S, Sahu G, Sinha A K and Saxena V K 2015 *Applied Thermal Engineering* **86** 222-228
[26] Chavan P D, Sharma T, Mall B K, Rajurkar B D, Tambe S S, Sharma B K and Kulkarni B D 2012 *Fuel* **93** 44-51
[27] Patil-Shinde V, Kulkarni T, Kulkarni R, Chavan P D, Sharma T, Sharma B K, Tambe S S, and Kulkarni B D 2014 *Industrial & Engineering Chemistry Research* **53** 18678-89
[28] Daggupati S, Mandapati R N, Mahajani S M, Ganesh A, Mathur D K, Sharma R K and Aghalayam P 2010 *Energy* **35** 2374-2386
[29] Daggupati S, Mandapati R N, Mahajani S M, Ganesh A, Sapru R K, Sharma R K and Aghalayam P 2011 *Energy* **36** 1776-1784
[30] Mandapati RN, Daggupati S, Mahajani S M, Aghalayam P, Sapru R K, Sharma R K and Ganesh A 2012 *Industrial & Engineering Chemistry Research* **51** 15041-15052
[31] Bhaskaran S, Ganesh A, Mahajani S, Aghalayam P, Sapru R K and Mathur D K 2013 *Fuel* **113** 837-843
[32] Sanga B, Mohanty D, Singh A K and Singh R K 2017 *Nexgen Technologies for Mining and*

Fuel Industries (New Delhi:Allied Publishers) p 1431-1436

[33] Sheng Y, Benderev A, Bukolska D, Eshiet K I, Gama C D, Gorka T, Green M, Hristov N, Katsimpardi I, Kempka T and Kortenski J 2016 *Mitigation and Adaptation Strategies for Global Change* **21** 595-627

[34] Olateju B and Kumar A 2013 *Applied Energy* **111** 428-440

[35] Eftekhari A A, Van Der Kooi H and Bruining H 2012 *Energy* **45** 729-745

[36] Cormos C 2012 *Energy* **42** 434-445

[37] Siefert N S and Litster S 2013 *Applied Energy* **107** 315-328

[38] Kunze C and Spliethoff H 2010 *Fuel Processing Technology* **91** 934-941

[39] Berkenpas M B, Frey H C, Fry J J, Kalagnanam J and Rubin E S 1999 *Integrated environmental control model technical documentation* (Pittsburgh: Center for Energy and Environmental Studies, Carnegie Mellon University)

[40] Singh U and Rao A B 2015 *Decision* **42** 191-209

[41] Singh U and Rao A B 2014 *International Conference on Advances in Energy Research* (IEEE) p 226–232

[42] Rao A B and Kumar P 2014 *Energy Procedia* **54** 431–438

[43] Chaulya S K et al 2016 *Development of Feasibility Assessment Model for Adaptation of Underground Coal Gasification Technology in North East Region of India* (submitted to Department of Electronics & Information Technology (DeitY), Ministry of Communication and Information Technology, Government of India, New Delhi)

Acknowledgement

AR and AKS are grateful to the CSIR Network Project *CoalGasUrja* (Development of Underground Coal Gasification Technology in India, ESC-0302) for financial support. Authors also thank Dr. Pradeep K Singh, Director, CSIR-CIMFR and Prof. D C Panigrahi, Director, IIT (ISM), Dhanbad for their skilful guidance and support. We thank the IECM team for making the software freely available. Further, thanks are due to Prof. Anand B Rao, IIT Bombay for his continued support in use of IECM software.

IOP Conf. Series: Earth and Environmental Science **76** (2017) 012008 doi:10.1088/1755-1315/76/1/012008

Computational Studies for Underground Coal Gasification (UCG) Process

Dipankar Chatterjee

Advanced Design and Analysis Group, CSIR-Central Mechanical Engineering Research Institute, Durgapur-713209, India

d_chatterjee@cmeri.res.in

Abstract. Underground coal gasification (UCG) is a well proven technology in order to access the coal lying either too deep underground, or is otherwise too costly to be extracted using the conventional mining methods. UCG product gas is commonly used as a chemical feedstock or as fuel for power generation. During the UCG process, a cavity is formed in the coal seam during its conversion to gaseous products. The cavity grows in a three-dimensional fashion as the gasification proceeds. The UCG process is indeed a result of several complex interactions of various geo-thermo-mechanical processes such as the fluid flow, heat and mass transfer, chemical reactions, water influx, thermo-mechanical failure, and other geological aspects. The rate of the growth of this cavity and its shape will have a significant impact on the gas flow patterns, chemical kinetics, temperature distributions, and finally the quality of the product gas. It has been observed that there is insufficient information available in the literature to provide clear insight into these issues. It leaves us with a great opportunity to investigate and explore the UCG process, both from the experimental as well as theoretical perspectives. In the development and exploration of new research, experiment is undoubtedly very important. However, due to the excessive cost involvement with experimentation it is not always recommended for the complicated process like UCG. Recently, with the advent of the high performance computational facilities it is quite possible to make alternative experimentation numerically of many physically involved problems using certain computational tools like CFD (computational fluid dynamics). In order to gain a comprehensive understanding of the underlying physical phenomena, modeling strategies have frequently been utilized for the UCG process. Keeping in view the above, the various modeling strategies commonly deployed for carrying out mathematical modeling of UCG process are described here in a concise manner. The available strategies are categorized in several groups and their salient features are discussed in order to have a good understanding of the underlying physical phenomena. This would likely to be a valuable documentation in order to understand the physical process of UCG and will pave to formulate new and involved modeling and simulation techniques for computationally modeling the UCG process.

1. Introduction

Computational modelling and simulation have become an inexpensive tool for predicting the influence of various physical and operating parameters on the performance of the UCG process, since conducting the UCG trials and data extraction is extremely costly and difficult. A lot of effort is being devoted to model the UCG process along with the field trials. The UCG process as shown in Figure 1 involves several complex mechanisms that happen simultaneously over a range of length and time scales. An integrated model involving several submodels is required in order to comprehensively describe the complicated UCG process. These submodels are:

- A coal submodel for investigating evaporation, pyrolysis, self-gasification and combustion of char, the effect of water influx, and heat and mass transport phenomena.
- A cavity submodel to address the cavity growth mechanisms, growth rate, cavity geometry, formation of the rubble pile at the bottom of the cavity, fluid flow, heat and mass transfer within the cavity.
- Gas cleaning and environmental submodels to address the separation of undesired gas species from the product gas, the surface treatment of the syngas for final use, and handling of the produced CO_2, H_2S, etc.

Combining all these sub-models would theoretically give an exact and complete description of the process; however building such a model is an extremely cumbersome task. Therefore all the previous models have mainly focused on studying these aspects in a decoupled manner through several simplifying assumptions.

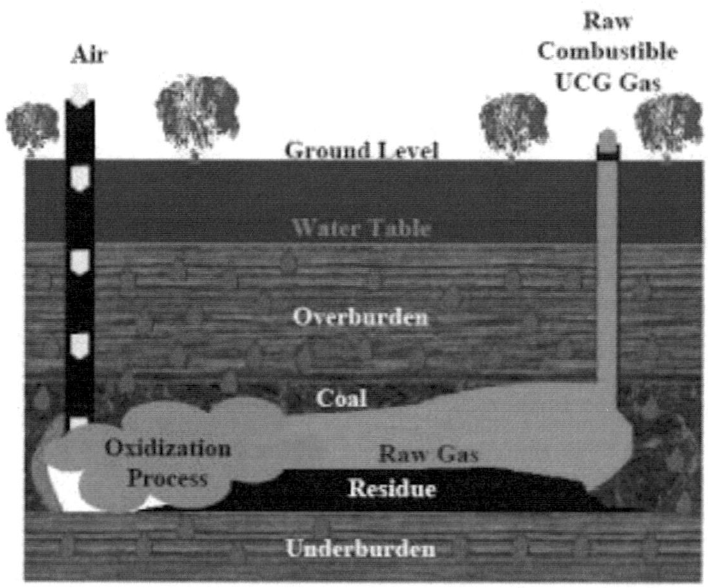

Figure 1. UCG process overview [1]

Current UCG models are mostly based on 1-D study of combustion and gasification of coal using a complex CFD (computational fluid dynamics) approach. These models are used to interpret and predict the performance of laboratory-scale experiments that are generally performed at lower pressures and oxidizing environments as well as a specific portion of field trials operating at pressures up to 5 MPa and shallow depths to 600 m. Some current UCG models assume complicated velocity equations, including the turbulent gas flow inside the cavity, and utilize chemical engineering correlations for a particle size and the porosity of the reactor. In addition, the cavity geometry and growth rate are modelled with an initially pre-assumed shape, either a tiny rectangular cube or a cylinder. Moreover, most of these studies focused on a single cavity and only a few considered interactions with the overburden. These models do not consider interactions between cavities. Due to the enormous complexity and computational cost of these models as the grid numbers increase, they cannot be efficiently used to simulate large-scale field trials. Using the CRIP technology, the coal is gasified in several stages and a sequence of cavities are developed according to the applied retraction strategies. These cavities are known to influence each other thermally. For this reason, the current models are not very appropriate to investigate the performance of different UCG technologies at a large scale.

The UCG models in the literature can be divided into two main categories; global and process models. Global models describe the overall field performance of UCG, while the process models are simple models of specific parts of the process, such as rock and coal spalling and the effect of water influx. The global models incorporate several approaches, such as packed bed, coal block, and channel models. In the following section, a brief overview of some important models will be provided and their approaches will be discussed in brief.

2. Overview of various models

2.1. Packed bed models

In the packed bed model, the process is described as a packed bed reactor. In this concept, the coal seam is considered to be a highly permeable porous media where the bed properties change with reactions. Winslow [2] used this approach to model the process in a 1-D domain with reactions occurring in char particles. Darcy equation was used to describe the fluid flow. Heat and mass transfer was addressed in the gas and solid phases, considering the convection terms only in the governing equations. Experimental correlations were used to describe the heat transfer between the phases. The numerical predictions were in accordance with the experimental results. Thorsness and Kang [3] extended the model by incorporating homogenous reactions, diffusion effects, wall transport, and also by revising the char reaction rates based on the ash segregation (AS) and shell progressive (SP) reaction models (refer to Figure 2).

(a) Shell progressive (SP) model (b) Ash segregation (AS) model

Figure 2. Different reaction regimes in the packed-bed model [2]

Khadse et al. [4] developed a simple 1-D model to describe the UCG process at char-rubble zone in a cylinder filled with coal particles. They followed a pseudo-steady approach, as there exists a large difference between solid and gas characteristic time-scales. Effect of O_2 concentration, Steam/O_2 ratio, and the inlet pressure were investigated. They found that increasing both O_2 and steam fraction in feed increases the propagation rate of the reaction front. Pressure was found to have insignificant impact. Although packed bed models are able to reproduce the laboratory experiments their extension to field trials is somewhat unfeasible, since other cavity growth mechanisms such as the thermo-mechanical failure cannot be incorporated into the model. Furthermore, this method requires fine grid system in the vicinity of the reaction front that practically limits its applicability.

2.2. Coal slab models

This concept considers the UCG coal seam as a coal slab. The process in these models is described by governing the movement of various defined regions in a coal slab. These regions include: gas film, ash layer, char region, dried coal, and virgin coal. There is an influx of the injection gases towards the cavity wall, while there is a simultaneous counter flux of steam and pyrolysis products from the wall to the cavity. The general framework of these models is depicted in Figure 3.

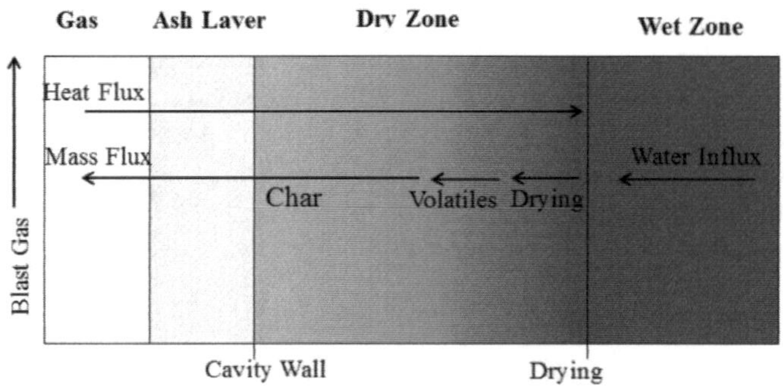

Figure 3. Scheme of the coal slab models [5]

Tsang [6] first used such approach and developed a 1-D unsteady model. The coal slab was divided into two zones; wet and dry zones. In the wet zone, heat conduction and liquid transfer are solved. It was assumed that all drying occurs in one moving point (Stefan model). This model neglects the effects of steam convection and assumes that the drying process is dominated by the conduction mode of heat transfer. In the dry region, pseudo-steady state assumption was adopted to solve the mass balance, heat transfer and gas-flux. Pyrolysis reactions are modelled as simultaneous independent reactions with kinetic parameters obtained from Campbell [7]. Porosity and permeability change with the extent of the reactions involving the correlations obtained from the oil well acidization. The model was validated with the experimental results of Forrester [8]. The model neglects combustion and provides the heat by putting a high temperature condition at the boundary. Massaquoi and Riggs [9-11] extended Tsnag's model [6] by incorporating the combustion of char and volatiles at the boundary. Three zones outside the coal slab were included: ash zone with a constant thickness, gas film, and the bulk gas. The model was used to describe simultaneous combustion and drying of a wet coal. The results showed that homogenous combustion of coal happen at lower surface temperature than reported for char particles. Furthermore, the burn rate decreases with increase in the concentration of O_2, when flame front is located in char face and increases when the flame is located in the gas film. Hong [12] applied previous submodels to develop a cavity model to predict the growth rate and composition of the final product gas. In this approach, the cavity was divided into several compartments representing a TIS model. Each compartment was assumed to include a void and a rubble pile region, representing the two branches of the parallel TIS model. The rubble pile (lower branch) and the void space (upper branch) were modelled using plug flow and continuously stirred tank reactors, respectively. It was assumed that all of the ash particles fall down on the rubble pile. This model was used to simulate the Hoe Creek III field trial. Park and Edgar [13] subsequently developed an unsteady 1-D model with a moving burning front to describe the lateral cavity growth in UCG. They also incorporated the coal shrinkage effect due to drying and pyrolysis, as well as steam and CO_2 gasification. Their results were validated with the experimental results of Poon et al. [14]. It was concluded that movement of the cavity wall due to shrinkage is only important in lab-scale process and can be neglected for larger scale process. Their results further indicate that the cavity growth is controlled by the rate of O_2 transfer to the cavity wall, when flame is located at the char surface. If flame is in the gas film, CO_2 and steam gasification determine the cavity growth rate. Abdel-Hadi and Hsu [15] extended previous coal block models by developing pseudo-2-D geometry with a moving burn front. Perkins and Sahajwalla [16, 17] used the CFD software FLUENT to describe UCG process in a 1-D coal block. This model is an extension of Tsang's model [6] through the inclusion of multi-component diffusion and random pore model to account for the change of heterogeneous reaction rates with conversion. They have proposed that cavity growth occurs at reducing conditions, therefore only heterogeneous gasification reactions are solved and required heat is provided by defining a constant temperature at the char surface. They assumed that solid and gas phases are in thermal equilibrium and bulk gas has a fixed composition that is representative of the product gas. Various experiments of pyrolysis and drying of large coal particles are used for model verification and an excellent agreement were reported. However, applicability of the model is limited and cannot be used to predict performance of UCG process because of the unphysical assumptions such as constant temperature at and fixed gas composition at the boundary. Results of their parametric study indicate that temperature of the coal surface, water influx and coal composition have the highest impact on the cavity growth rate. Nourizadeh et al. [18] developed a 3-D model using the reservoir simulation software CMG-STARS based on the finite-difference method. They have introduced an arbitrary clay layer in the seam to investigate effect of inert layers on the process; however thermo-mechanical failure of this layer is not included in the model. Chemical reactions and conduction/convection of heat and species were included in the model. The results reveal a large degree of grid dependence and not validated properly. Further, the model was extended to simulate CRIP with successive ignition points [19]. However, again no validation or comparison with the experimental results was reported.

2.3. Channel models

The UCG process, in this approach, is represented by an expanding channel when two distinct zones of rubble/char and open channel exist. Channel regime could dominate the process at a later stage. Figure 4 shows the basic concept and the underlying physics [17]. Air or O_2 flows down the central channel and is transported by turbulent flow to the boundary layer along the channel wall. The O_2 diffuses through the boundary layer to the solid surface and reacts. The hot combustion gases diffuse back through the boundary layer to the channel. Magnani [20] developed a 1-D channel model with a fixed diameter, assuming that neither axial nor radial diffusion exists, and there are constant axial convection, two reactions of char oxidation and Boudouard reactions, constant reaction rates, and only axial heat transfer in the solid phase. This model was used to investigate gas composition and temperature profile trends along the channel. Pasha and Farouq-Ali [21] improved the shortcomings of Magnani [20] by including radial and axial diffusion and heat transfer in the solid phase, variable gas velocity and channel diameter.

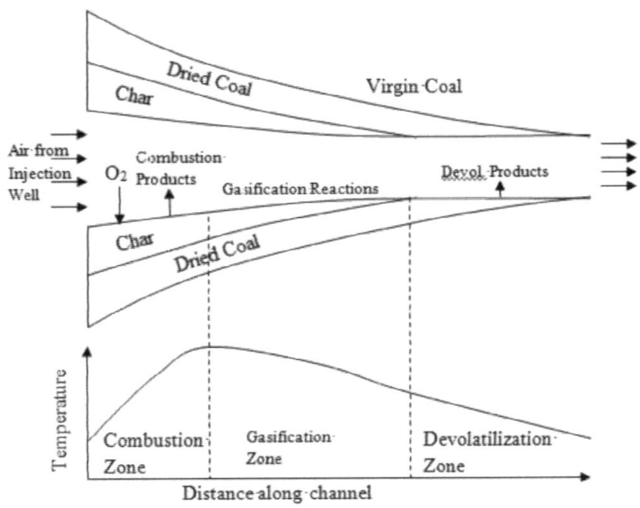

Figure 4. Reactions and Transport phenomena in channel model [17]

Dinsmor et al. [22] developed a steady-state 2-D model by assuming that the gasifier is a cylindrical channel with reactions occurring at the walls. Mass transfer correlations were calculated based on the turbulent flow in pipes. It was concluded that the operation of UCG in the channel regime is undesirable and would lead to a decrease in the gas heating value due to the bypass of O_2. In this model, larger channels would lead to lower velocity and Reynolds number, which in turn decreases the mass transfer by forced convection. The growth rates obtained from simulation were found lower than the earlier field trials. However, the model was able to highlight the importance of natural convection in channels, which increases the heat and mass transfer. The effect of natural convection in cavity growth was first considered by Eddy and Shwartz [23] through a 2-D model. They described the evolution of the cavity based on the movement of the cavity wall. All reactions are assumed to occur at the wall under a constant temperature condition. Transport of heat and species inside the cavity is governed by empirical correlations. The results indicated that the natural convection is several orders of magnitude higher than the forced convection. The model was able to reproduce the results of Hanna II and Pricetown field trials qualitatively. Later, Luo et al. [24] extended the model by including the heat transfer and coal wall reactions. Flow inside the cavity is solved following the irrotational fluid flow inside an enclosure (Figure 5). FLUENT was used to predict the cavity shape in different times based on the heterogeneous reaction rates at the cavity wall. The model was validated with Chinchilla and Hanna II results [24]. This model predicts hemispherical cavity geometry, however, its use is limited since heat and mass transfer characteristics are unknown. Also, the coupling of this model with the mechanical failure of coal would be cumbersome as the accumulated rubble changes the transport characteristics inside the cavity.

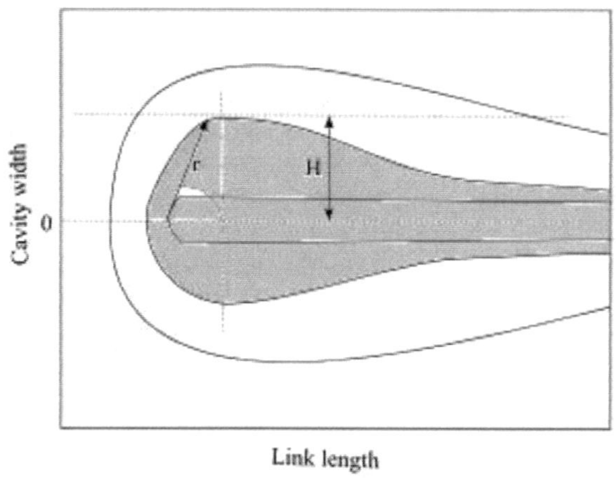

Figure 5. UCG cavity defined in the work of Luo et al. [24]

At thin seams, failure of the overburden has a major effect on the process. Such a failure would create open channels; as shown in Figure 6. Various models have so far been developed to describe the process in these channels.

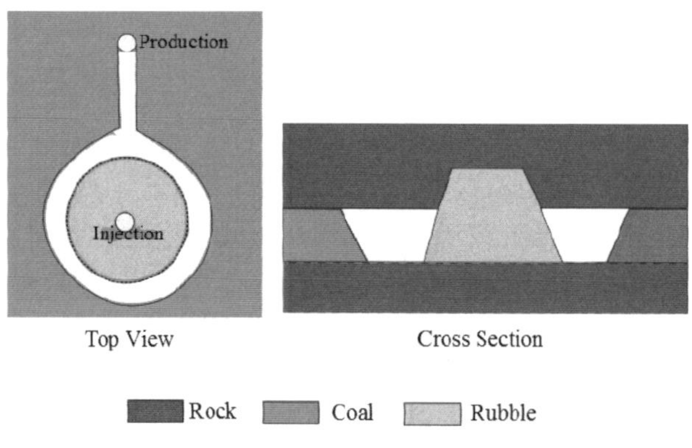

Figure 6. Channel formation in thin seams [28]

Pirlot and co-workers [25, 26] developed a 1-D steady-state channel model with two distinct zones: high and low permeability rubble. Heat and mass transfer were calculated from empirical equations of packed bed concept. Reaction rates of the gas phase were written for an ideal plug flow. Coal consumption rate was calculated on the channel wall. The model was further extended to 2-D to simulate the cavity growth and shape. Batenburg et al. [27] developed a steady 2-D model for UCG in open channels. Reaction rates were calculated based on resistances in the system including boundary layer, pore diffusion, surface phenomena and chemical kinetics. Heat transfer was addressed by radiation between walls of channel. The effect of natural convection was also included in their model. Results indicated that natural and forced convection are equally important. Kuyper [28, 29] developed a 2-D model (Figure 7) to describe the UCG process in a cross-section of an open channel. The *k*-*ε* turbulence model was used to describe the fluid flow and the heat transfer was governed by radiation and convection. Results showed that O_2 is consumed far from coal wall by combustible gases. Double-diffusive natural convection would cause periodic generation and collapse of CO_2 bubbles. Also, mass transfer is reported to be the controlling mechanism for reduction reactions at coal wall. They also investigated the effect of CO_2 injection into the coal seam and reported that it has similar effect as adding steam; although to a lesser extent.

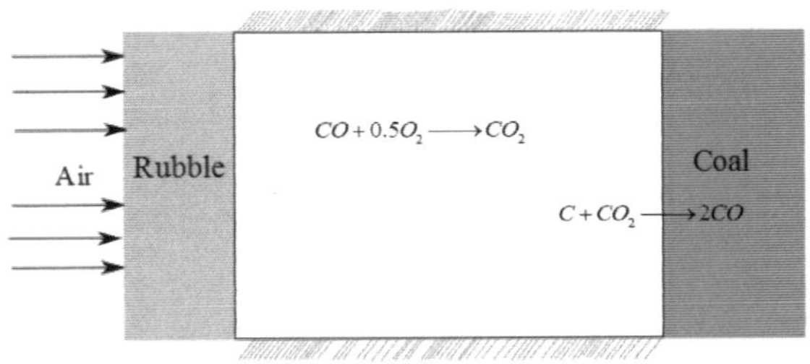

Figure 7. Channel geometry in Kupyer's model [29]

Perkins and Sahajwalla [30] extended Kuyper's model [28, 29] by including an ash layer at the lower part of the channel. The axi-symmetric model is used to investigate natural convection along with relevant reactions in a partially filled cavity (Figure 8). Flow was assumed to be laminar in the ash zone and turbulent in the void space. They concluded that O_2 should be injected at the bottom of the channel, otherwise, valuable gasification products would be oxidized leading to low heating value of the production gas.

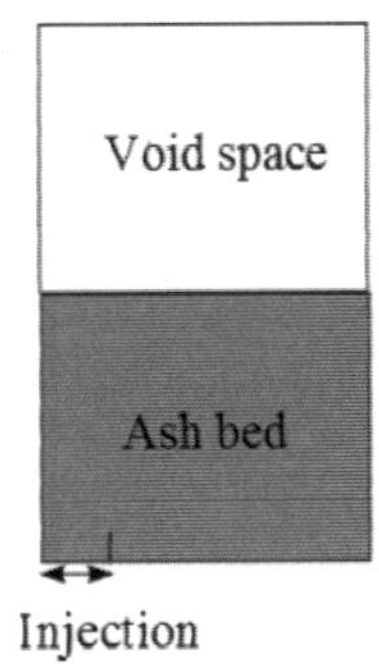

Figure 8. Channel geometry in Perkins and Sahajwalla [30]

2.4. Other approaches for UCG modelling

2.4.1. Reactor models

In this approach, the UCG process is described as a series of ideal reactors. Such models can be used to predict product gas composition without considering the complex phenomena occurring underground. Chang et al. [31] used this approach to predict the resource recovery during UCG process. In this model, the domain is divided into cylindrical elements, where each element consists of two distinct zones: rubble and void. It is assumed that the rubble zone acts as an ideal Plug Flow Reactor (PFR), while void space is assumed to be perfectly mixed modelled as a CSTR. The spalling and water influx are calculated based on the empirical correlations. The results revealed the importance of turbulent transport in field trials. Some researchers have used the response of tracer material to characterize various zones in the underground cavity and track the cavity growth. Based on the residence time of tracer elements, cavity is modelled as an arrangement of ideal reactors. Debelle et al. [32] used different chemical engineering reaction models to fit the quasi-exponential decay of tracer response. Different stages of the process such as reverse combustion linking and forward gasification were represented through regression models. For gasification period, it was observed that cavity offers a high level of back-mixing, resembling a well-mixed reactor with a small dead zone. Thorsness [33] performed tracer experiments to explore the extent of water influx and steam injection requirements. Pirard et al. [34] used helium tracer and found that cavity behaves almost like a small number of CSTRs in series with high level of back mixing. Daggupati et al. [35] described the cavity as series of ideal reactors (Figure 9). Residence Time Distribution (RTD) of gases was calculated based on the CFD results. This approach has been validated by laboratory-scale tracer experiments. They summarized that as the cavity size increases, cavity behaviour changes from PFR to CSTR.

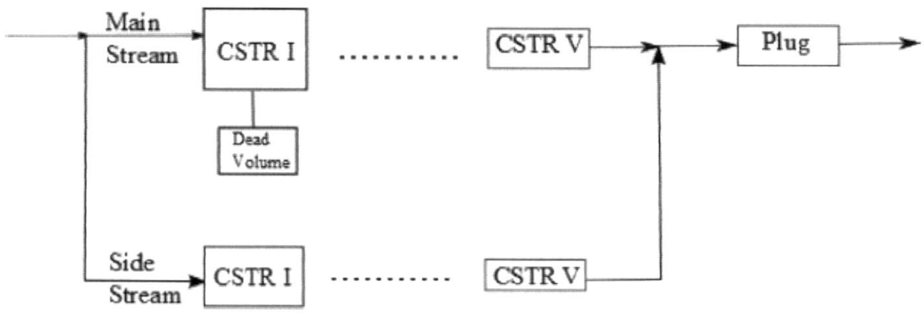

Figure 9. Reactor arrangement for cavity in works of Daggupati et al. [35]

2.4.2. Probabilistic simulation

Batenburg and Bruining [36] adopted probabilistic simulation to describe forward gasification in 2-D UCG geometry. The model uses stream function on the grid blocks. The derivative of the stream function along ash-char interface is interpreted as the possibility that interface is moving in a certain block. The applicability of this model is very limited since the temperature of the gases and composition are considered constant in each region. Biezen et al. [37] extended the concept by including movement of several interfaces including ash-void, ash-coal, void-coal, and void-rock (Figure 10). The model consists of two modules: one module solves flow in the entire domain and the other selects a block of coal/rock for gasification or spalling. A single streamline is selected randomly and based on the interface located on the streamline, relevant physical phenomena proceeds. They later extended their model to simulate cavity evolution in a 3-D domain [38]. This model is used to simulate the development of gasifier from early times to a developed stage. Application of this model is limited due to its complex nature and certain unphysical assumptions such as constant temperature at each domain.

2.4.3. Process models

All of these reviewed modelling efforts so far have the deficiency of focusing on some aspect of the UCG process, while neglecting others. During UCG trials in US, LLNL funded many modelling studies for UCG process. Several 1-D models were developed and validated with certain series of data from field trials. These segregated models were later combined together to form the CAVISM process model that represents 15 years of continuous UCG research and development in US [39]. The model is applicable for predicting lateral cavity growth of thick and shrinking coal seams in which O_2 is injected at the bottom of the coal seam. The model is based on some major assumptions such as:

cavity is axisymmetric around injection; thermal radiation is the main heat transfer mechanism in well-mixed void space; cavity growth is dominated by thermo-mechanical failure of wall and a packed bed of char and rubble forms over a thin layer of ash. Various sub-models were added into this model to quantify water influx, flow dispersion through a rubble bed, sidewall growth of cavity due to reactions, and spalling of cavity roof into the domain. This model has reproduced results of some of the field trials in US. Therefore, the model could not be used in other seams, as these parameters are not known *a priori*.

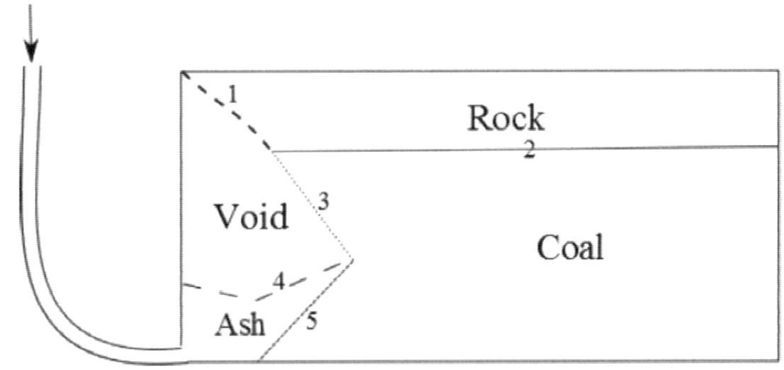

Figure 10. Probabilistic simulation of UCG process by possibility of movement of following interfaces: 1.Void-Rock; 2. Coal-Rock; 3. Void-Coal; 4.Void-Ash; 5. Ash-Void. [38]

3. UCG studies in India

India has a vast coal reserve which are the fourth-largest in the world, primarily of low grade. UCG may be a very promising technology for such coal reserves in India, to reduce the dependency on the oil and gas imports. UCG is expected to be used to utilize India's large coal reserves, which are difficult to extract economically using the existing conventional technologies. The Government of India constituted a National Committee on UCG taking experts from the University of Bombay, Ministry of Petroleum, Planning Commission, ONGC, CIL/CMPDI, CIMFR (the then CMRS and CFRI), ABE, EIL and IIT Kharagpur in July 1984. A protocol for UCG development was also signed between the Government of India and the Government of erstwhile USSR in 1981. It was also decided that UCG prospect evaluation for coal seams of less than 300m depth would be handled by ONGC. Gautam and Singh [40] have presented a summary of Indian efforts.

As UCG technology for intermediate depths had already been experimented and commercialized in other countries, therefore, pilot project in India was designed to acquire relevant technology at greater depths. Area chosen was near Mehsana city in North Gujarat province where large coal reserves of about 63 billion tonnes at depths ranging from 700 to 1700 m exist. The study covered an area of about 6.7 sq km having an average thickness of coal seam of 15 m. The recoverable energy from these reserves in the form of gas was estimated to be equivalent to 15,000 BCM of natural gas. ONGC drilled 2 pilot wells UCG-1 and UCG-3 for evaluation of the rock characteristics, sub-surface strata conditions and quality of coal. Coal cores retrieved were analysed for various parameters. Of the 15 coal horizons encountered, 11 coal horizons were of thickness more than 1m occurring between 745-941 m of depth. The cores recovered from a depth of about 852 m were highly fractured. Oil traces were observed along these fractures. The coals were high moisture, high volatile, low rank non-caking lignitous type having carbon content (dry ash free basis) between 72 and 76% but somewhat higher hydrogen content compared to other lignite deposits in India. Appreciable quantities of methane content varying from 1 to 6 m3/t were observed in the coal cores. Keeping in view the geohydrological conditions and roof and floor characteristics, Shobhasan-III seam was identified as the most suitable coal seam for in-situ gasification. However, further studies on the subject have not been made since then. Recently ONGC has signed an MOU with Stochinsky Institute of Mining, Russia for taking up UCG projects in India. ONGC has also signed an MOU with Coal India Limited and another with the Singareni Collieries Company Limited. It is likely that ONGC would very soon enter into an MOU with Gujarat State Petroleum Corporation for cooperation in developing and applying the UCG technology. The Ministry of Coal, Government of India, has awarded an S&T study project to the Neyveli Lignite Corporation.

Recently, CFD based studies of complex flow patterns in a developing UCG cavity were conducted by researchers from IIT-Bombay in collaboration with ONGC [41]. The main objective of this work was to understand the velocity distribution and perform residence time distribution (RTD) studies in the UCG cavity. Based on the RTD studies, the actual UCG cavity at different times was modeled as a simplified network of ideal reactors, which might offer a computationally less expensive and easier option to determine UCG process performance as a function of time. In another effort, Chatterjee and co-workers [42] conducted a comprehensive three-dimensional numerical study to understand the hydrodynamics within a given cavity size which would give us a relatively quick but reliable insights into the process. Five different cavity sizes were considered inside which the complete turbulent transport was simulated. Apart from the usual vertical and horizontal injection, the effect of inclined injection on the hydrodynamics was also reported in their study for the first time. The numerical simulation was performed by using a commercial CFD package Ansys Fluent [44].

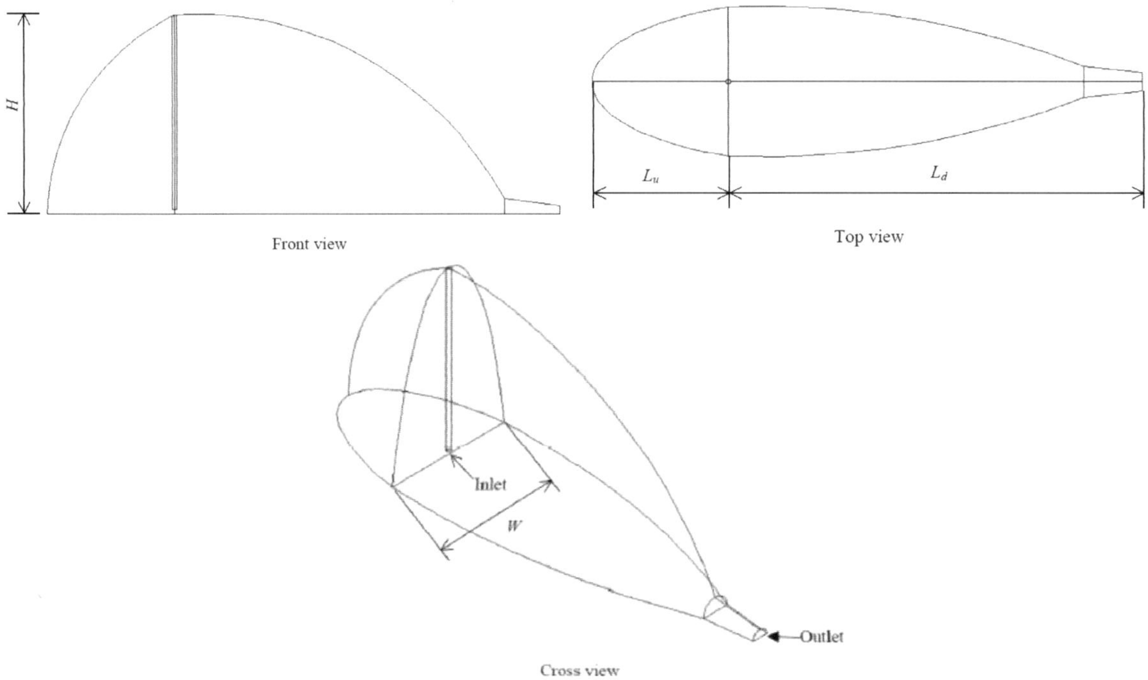

Figure 11. Geometry of UCG cavity in [42]

4. Summary

Although several numerical models involving varying levels of complexities have been published over the past years, applicability of these models is limited to some specific and isolated cases only and the models could hardly be generalized to predict the performance of the practical UCG trials. Furthermore, most of these models are in the 1-D or 2-D framework. Even the 3D models developed thus far are very simple and did not consider the details of the physical and chemical models. Also, the fluid flow is laminar in early stages of the process while changes to turbulent in a developed channel. Turbulence would arise from natural convection due to thermal and solutal gradients in the cavity. Developed models are meant to simulate one of these stages and could not be used in these two conditions. Hence, the future focus should be dedicated towards developing a model in a 3-D framework including all the factors affecting the performance of the UCG and can correctly reproduce the practical UCG field trials.

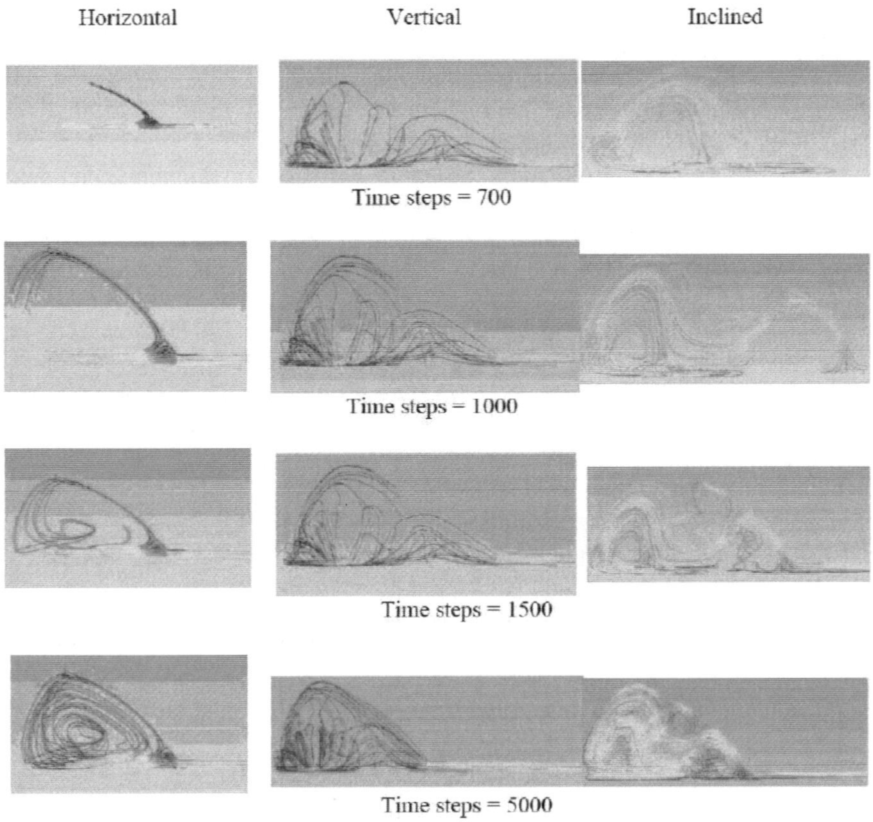

Figure 12. Flow dynamics (pathlines) colored by velocity magnitude (in m/s) at different time steps and different nozzle orientations in [42]

References

[1] http://coalandfuel.blogspot.in/2008/09/underground-coal-gasification-ucg.html
[2] Winslow A M 1976 *Numerical model of coal gasification in a packed bed* (UCRL-77627, LLNL Report, Berkeley, CA)
[3] Thorsness C B and Kang S W 1985 *Further development of a general-purpose packed bed mode for analysis of UCG process* (Report no. UCRL-92489, LLNL Report, Berkeley, CA)
[4] Khadse A N, Qayyumi M, Mahajani S and Aghalayam P 2006 *Int. J. Chem. Reactor Eng.* **4** 1
[5] Gunn R D and Krantz W B 1987 *Underground coal gasification: development of theory, Laboratory Experimentation, Interpretation, & Correlation with the Hanna Field Tests: Final report* (U.S. Department of Energy: Morgantown, WV, USA)
[6] Tsang T H T 1980 *Modeling of heat and mass transfer during coal block gasification* (Ph.D. Thesis, University of Texas at Austin, Austin, TX)
[7] Campbell J H 1978 *Fuel* **57** 217
[8] Forrester R C 1979 *5th Annual Underground Coal Conversion Symposium* (U.S. Department of Energy: Washington, DC) p 402
[9] Massaquoi J G M and Riggs J B 1983 *Chem. Eng. Sci.* **38** 1747
[10] Massaquoi J G M and Riggs J B 1983 *Chem. Eng. Sci.* **38** 1757
[11] Massaquoi J G M and Riggs J B 1983 *AIChE J.* **29** 975

[12] Hong L C 1984 *Process analysis and simulation of underground coal gasification* (PhD Dissertation, The University of Texas at Austin, Austin, TX)
[13] Park K Y and Edgar T F 1987 *Ind. Eng. Chem. Res.* **6** 237
[14] Poon S S M, Edgar T F and Park K Y 1986 *In Situ* **10** 109
[15] Abdel-Hadi Eed A A and Hsu T R 1987 *J. Energy Resour. Technol.* **109** 11
[16] Perkins G and Sahajwalla V 2005 *Energy Fuels* **19** 1679
[17] Perkins G and Sahajwalla V 2006 *Energy Fuels* **20** 596
[18] Nourozieh H, Kariznovi M, Chen Z and Abedi J 2010 *Energy Fuels* **24** 3540
[19] Seifi M, Chen Z and Abedi J 2011 *Can. J. Chem. Eng.* **89** 1528
[20] Magnani C F 1973*Mathematical foundations for the process of underground gasification of coal* (PhD Dissertation, The Pennsylvania State University)
[21] Pasha M L and Farouq-Ali S M 1979 *14th Intersociety Energy Conversion Engineering Conference* (Boston) vol. 1. (A79-51726 23-44) (Washington, D.C., American Chemical Society) p 1010
[22] Dinsmoor B, Galland J M and Edgar T F 1978 *J. Pet. Technol.* **30** 695
[23] Eddy T L and Schwartz S H 1983 *J. Energy Resour. Technol.* **105** 145
[24] Luo Y, Coertzen M and Dumble S 2009 *7th International Conference on CFD in the Minerals and Process Industries* (CSIRO, Melbourne, Australia)
[25] Mathy B et al. 1994 *In Situ* 18 399
[26] Pirlot P, Pirard J P, Coeme A and Mostade M 1998 *In Situ* **22** 141
[27] Batenburg W D, Biezen N J and Bruining J 1994 *In Situ* **18** 419
[28] Kuyper R A, van der Meer T H and Hoogendoorn C J 1994 *Chem. Eng. Sci.* **49** 851
[29] Kuyper R A, van der Meer T H and Bruining J 1996 *In Situ* **20** 311
[30] Perkins G and Sahajwalla V 2007 *Chem. Eng. Res. Des.* **85** 329
[31] Chang H L, Himmelblau D M and Edgar T F 1985 *In Situ* **9** 185
[32] Debelle B, Malmendier M, Mostade M and Pirard J P 1992 *Fuel* **71** 95
[33] Thorsness C B 1980 *Steam tracer experiment at the hoe creek no.3 underground coal gasification field test* (Report UCRL-53082, LLNL, CA)
[34] Pirard J P, Brasseur A, Coeme A, Mostade M and Pirlot P 2000 *Fuel* **79** 471
[35] Daggupati S et al., 2011 *Ind. Eng,. Chem. Res.* **50** 277
[36] Batenburg W D and Bruining J 1993 *In Situ* **17** 413
[37] Biezen E N J, Bruining J and Molenaar J 1994 *Annual technical conference of the Society of Petroleum Engineers* (New Orleans, LA, USA)
[38] Biezen E N J, Bruining J and Molenaar J 1995 *Annual technical conference of the Society of Petroleum Engineers* (Dallas, TX, USA)
[39] Britten J A and Thorsness C B 1989 *In Situ* **13** 1
[40] Gautam N N and Singh A K 2005 Proc. of the Coal Summit (New Delhi) p 207
[41] Aghalayam P, Sateesh D, Naidu R, Mahajani S, Ganesh A, Sapru R K and Sharma R K 2008 25th Annual International Pittsburgh Coal Conference (Pittsburgh, PA) Paper 32-4
[42] Chatterjee D, Gupta S K, Aravind C and Roshan R 2014 *J. Adv. Thermal Sci. Res.* **1** 15

Acknowledgments
The materials in this review paper are taken from several published works. The author sincerely acknowledges all these contributions. Special acknowledgement to Ahad Sarraf Shirazi (MS Thesis, University of Alberta, 2012) and Mojtaba Seifi (PhD Thesis, University of Calgary, 2014) in this regard. Financial support from CSIR (Govt. of India) through network project (no. ESC0302/05) is gratefully acknowledged.

IOP Conf. Series: Earth and Environmental Science **76** (2017) 012009 doi:10.1088/1755-1315/76/1/012009

Application of the Exergy UCG technology in international UCG projects*

M S Blinderman

Ergo Exergy Technologies Inc., Montreal, Quebec, H2Y2R6, Canada
Email: michael.blinderman@ergoexergy.com

Abstract. Underground Coal Gasification is a subject of continuing global interest in the energy sector. While the international scenario in UCG is promising, it is deeply desirable that advances in this area are seen in India as well. This is particularly so with the Paris Climate Agreement bringing in more stringent challenges for clean energy development. India has many potential coal basins which may be suitable for UCG deployment. India is in dire need of indigenous source of gaseous and liquid hydrocarbons that could compete with imported products. It is also the country with exceptionally large and diverse coal and lignite resources, large part of which could not be mined due to geological complexity and prohibitive cost. Thus, there is a rationale that the εUCG™ technology plays a decisive role in realizing the potential of Indian coal resources for the benefit of Indian industry and population.

1. Background

Underground coal gasification (UCG) can be commercialized only if it demonstrates consistent syngas quality and consistent production output (with a variability of less than 1%) and can be applicable at a commercially significant scale. What is a commercial scale for a UCG plant? We would like to suggest that UCG should operate in a cost-effective and environmentally sustainable way, producing syngas flow rates equivalent to about 1 Mt/year of coal or lignite, with an economic plant life of up to 30 years. (Although in some special circumstances, smaller plants may also proof to be commercially viable). As an example, a 300 MW IGCC power plant running on syngas produced by a εUCG™ plant gasifying typical Gujarat lignite would consume about 1.7 MT of coal per annum. Thus, commercial plant requires large coal reserves of the order of 50 million tons, with extraction efficiency of 90-95%, which has been demonstrated by εUCG™ technology. In terms of stable quality of syngas product, the following process represents an inherent challenge for UCG technologies. The UCG cavity can be considered as a chemical reactor with walls made up of coal. This means that the size of this reactor is continuously changing, as the diameter goes on increasing. This leads to lower gas velocity due to the equation of continuity. A lower gas velocity leads to less turbulence, which leads to a poorer level of heat and mass transfer. This, therefore, leads to lowering gas quality eventually showing increased oxygen content in the production well. At this point of time, the operations need to be stopped and parameters revised. Any successful UCG technology aspiring to achieve commercialization must demonstrated how it overcomes this syngas deterioration phenomenon.

Controlled retracting injection point (CRIP) has been suggested by Hill and Shannon [1] as a solution for syngas deterioration issue. This allows the operator to choose optimal time and distance in movement of the injection point. It is a very popular approach and many simulation studies have been developed to model this process [2]. However, the process has inherent issuess as the gas quality

* This article has been adapted by Dr. Ajay K. Singh from a lecture delivered at the "Workshop on Challenges and Opportunities of Underground Coal Gasification", Vigyan Bhawan, New Delhi on 14 February 2017.

deteriorated in the same way as conventional UCG processes (even though the quality was slightly better) as evidenced by performance of Rocky Mountain 1 UCG trial (1986-1988). As an alternative to this, parallel CRIP method was suggested. This also presents some in-built challenges as it turns uneconomical with greater width of the channel (the effect of "cannibalizing" the product gas). This problem may be solved by allowing the roof to collapse strategically such that the width of the channel reduces to around 3 metres and improvement of reaction is seen. Therefore, the basic principle here is that the cavity should not be allowed to grow uncontrollably, failing which the reaction process becomes extremely slow. Another issue with CRIP based systems is the mine plan i.e. significant cross-flow and mixing of the gases in the reactors was seen if they were not far apart. With increase in the distance between the two reactors, the extraction efficiency reduces to uneconomic levels.

Similarly, bord and pillar method may also not be used to UCG processes due to risks of spontaneous combustion, uncontrolled overburden deformation and loss of water resources. Thus, there may be a great difficulty in completely rehabilitating the operational site.

2. The εUCG™ technology

The Exergy UCG™ (εUCG™) technology, unlike a conventional UCG processes, is not a two well process. Rather, it is a large scale mining method. It has inbuilt advantages in terms of rock deformation and water influx management. It involves drilling of directional, inclined, vertical and other wells, and has been developed based on 75 years of Soviet UCG experience. The average panel capacity in this technology is about 5 PJ/a, while the mine average coal extraction rate is around 95%. The technology has shown suitability with a wide range of coals as shown in Table 1.

Table 1. Deployment of εUCG™ technology in various UCG projects

UCG Plant	Rank	Thickness, m	Depth, m	Dip, °	LHV, MJ/kg
Lisichansk	Bituminous	0.44 - 2.0	60-250	38-60	20.1-23.0
Yuzhno-Abinsk	Bituminous	2.2-9.0	130-380	35-58	28.9-30.7
Podmoskovnaya	Lignite	2.5	30-80	<1	11.8
Angren	Lignite	3.0-24.0	110-250	7	15.3
Shatskaya	Lignite	2.6	30-60	<1	11.0
Sinelnikovo	Lignite	3.5-6.0	80	<1	8.0
Chinchilla	Sub-bituminous	10.0	135	<1	21.7
Majuba	Bituminous	3.5-4.5	285	3	20.3
Kingaroy	Sub-bituminous	17.0	200	5	23.5
Huntly West	Bituminous	4.0-22.0	220-540	0-75	24.5
CC Alberta	Sub-bituminous	7.0	150-260	6	20.5-23.0
Alaska SHR	Lignite/Sub-bituminous	1.0-12.0	50-650	0-75	11.0-16.5

The Exergy UCG™ (εUCG™) technology is a technology for extracting the energy and hydrocarbons from underground unminable fossil fuels deposits, such as coal, lignite or shale, by converting them into gaseous and liquid hydrocarbons in a gasification process performed in situ, whereby drilled wells are used for delivering oxidant for gasification process and transporting products to the surface for processing and beneficial use. The εUCG™ technology was developed by Ergo Exergy Technologies Inc and has been applied in multiple international UCG projects.

The εUCG™ technology can help create a large-scale source hydrocarbons from unmineable coal or lignite in the countries and areas where there is no other source of natural gas and petroleum. This source of hydrocarbons is indigenous, safe, environmentally clean and carbon-efficient. Thus produced, the hydrocarbons have highly competitive costs and therefore can be used as fuel for high-efficiency power generation or as raw materials for synthesis of clean fuels and chemicals, including automotive fuels, ammonia-based fertilizers, synthetic methane, DME etc.

The εUCG™ technology is a large-scale mining technology with a typical panel capacity of about 0.3 million metric tonne per annum. The process incorporates controlled rock deformation and ground water influx into underground reactors. Depending on quality of coal and specifications for required products, the process may include injection of compressed air, oxygen, steam and water, carbon dioxide and other oxidants. Injection of oxidants and production of syngas is effected via drilled boreholes, design of which is suited to geologic condition and includes vertical, inclined and directional wells. In the root of the εUCG™ technology lies the stream method of UCG, developed and commercially exploited in the former USSR for more than 70 years. In the last 20 years the εUCG™ technology has been applied in the projects developed in Australia, New Zealand, Canada, South Africa and the USA.

In relation to global warming concerns, the gaseous products of εUCG™ technology can be used for power generation in a high-efficiency IGCC plants with significant reduction of CO_2 emissions compared to coal-fired plants. Besides, the concentrated gaseous product stream of the εUCG™ technology provides an easy and inexpensive opportunity for CO_2 capture and sequestration. The latter is aided by creation of underground cavities in the course of gasification process, which after cooling may be used for storage and sequestration of supercritical carbon dioxide. In this case the carbon footprint of a εUCG™ IGCC plant would be lower than that of a natural gas CCGT plant. The typical cost of hydrocarbon products of the εUCG™ technology is approximately $1.00 per MMBTU, although the actual cost may vary with the depth and thickness of the coal seam and the coal quality. Specific projects where εUCG™ technology has been applied will be considered in this lecture in detail. They took place in greatly varying geologic and hydrogeological conditions and have been very successful in demonstrating capabilities of technology to safely produce low cost hydrocarbons while always protecting the environment, especially the groundwater.

It is expected that this technology will lead to significant cost reductions in Indian conditions(as shown in Table 2).

Table 2. Costs of various product with εUCG™ technology, as against market price

Parameter	εUCG™	Market price
Cost of Electricity ($/MWh)	21.79	48.23
Cost of SNG ($/mmBTU)	2.14	2.72
Cost of Methanol ($/tonne)	69	315
Cost of Gasoline ($/Gallon)	0.66	3.90
Cost of Diesel ($/Gallon)	0.94	3.22
Cost of Urea ($/tonne)	106	413

3. Conclusions

There is significant repository of resource, regulations and expertise in terms of UCG deployment in India. The εUCG™ is a well-established technology with evident advantages, already shown in the Chinchilla I project and other operations around the world. The technology overcomes several of the shortcomings shown in the CRIP and bord and pillar methods. The cost of various commodities with this technology is well below the market price. The technology has also demonstrated sustainability as no major environmental issues have surfaced in the operations [3]. The writing on the wall is now clear – UCG in India needs to take-off with committed developers and investors. Coalfields in various part of the country may be suitable for UCG and an optimum path may be set for its deployment to generate a clean energy alternative, which also caters to natural gas availability and cheap electricity.

References

[1] Hill R W and Shannon M J 1981 *Proceedings of the Seventh Underground Coal Conversion Symposium* (Fallen Lake) LLNL p 730-737

[2] Seifi M, Chen Z and Abedi J 2011 *The Canadian Journal of Chemical Engineering* **89** 1528-1535

[3] Blinderman M S 2006 *ASME 2006 Power Conference* (Atlanta: ASME) p 437-444

IOP Conf. Series: Earth and Environmental Science **76** (2017) 012010 doi:10.1088/1755-1315/76/1/012010

An overview of the geological controls in underground coal gasification

Debadutta Mohanty

Non-Conventional Gases Section, CSIR-Central Institute of Mining and Fuel Research, Dhanbad 826 015, India

Email: drdmohanty@ymail.com

Abstract. Coal's reign will extend well into this millennium as the global demand for coal is expected to increase on average by 2·1% per year through 2019. Enhanced utilization of the domestic coal resource through clean coal technologies is necessary to meet the energy needs while achieving reduced emissions. Underground coal gasification (UCG) is one of such potential technologies. Geology of the area plays decisive role throughout the life of a UCG project and imperative for every phase of the project cycle starting from planning, site selection, design to cessation of operations and restoration of the site. Impermeable over/underlying strata with low porosity and less deformation are most suitable for UCG processes as they act as seal between the coal seam and the surrounding aquifers while limiting the degree of subsidence. Inrush of excess water into the gasification chamber reduces the efficacy of the process and may even quench the reactions in progress. Presence of fresh water aquifer in the vicinity of target coal seam should be abandoned in order to avoid groundwater contamination. UCG is not a proven technology that is still evolving and there are risks that need to be monitored and managed. Effective shutdown programme should intend at minimising the post-burn contaminant generation by flushing out potential organic and inorganic contaminants from the underground strata and treating contaminants, and to restore ground water quality to near baseline conditions.

1. Introduction

Coal is the major contributor to electricity generation holding a worldwide market share of 40 per cent being mainly driven by growing economies. The increasing contribution of renewable energy sources is not just sufficient to meet the ever increasing energy demands of the society and, coal and other fossil fuels will continue to be a major source of energy in the future. Coal's reign will extend well into this millennium as the global demand for coal is expected to increase on average by 2·1% per year through 2019 [1]. India is the world's third largest coal producer. The 2015 Indian coal inventory estimated, including only seams greater than 0.9 m in thickness and up to a depth of 1200 m, at 306.6 billion metric tons out of which less than 100 billion metric tons are recoverable by conventional mining methods. Besides hard coal, India is endowed with 44.1 billion metric tons of lignite resource. While India has vast coal resources, it also has huge petroleum and natural gas deficits. In such a scenario it is difficult to achieve the emission reduction targets from fossil fuels by using conventional coal-based technologies. Enhanced utilization of the domestic coal resource through clean coal technologies is necessary to achieve energy security while achieving reduced emissions (e.g. particulates, NO_x, SO_x and CO_2). Underground coal gasification (UCG) is one of such potential technologies to meet the future energy needs of coal-bearing countries worldwide in a cleaner manner

[2-4]. UCG technology can be used in India to produce clean energy using its domestic coal resources, including currently unmineable deposits, at competitive costs and with minimum damage to the environment. UCG is not a proven technology that is still evolving and there are risks that need to be monitored and managed.

About fifty UCG trials (Figure 1) were made in various countries including FSU, United States, Canada, Belgo-German, China, South Africa and Australia etc. most of which were short lived except Angren project of Uzbekistan which is operating over 50 years [5]. The trials in various countries, however, showed that UCG is highly adaptive to different conditions such as horizontal to steeply dipping coal seams of varying ranks, depths, geology and hydrogeological etc. New modern technologies with the upstream oil and gas industries along with century long trials provided further thrust to UCG to beyond the proof-of-concept phase to a phase of commercialization that is solely guided by the proper site characterization.

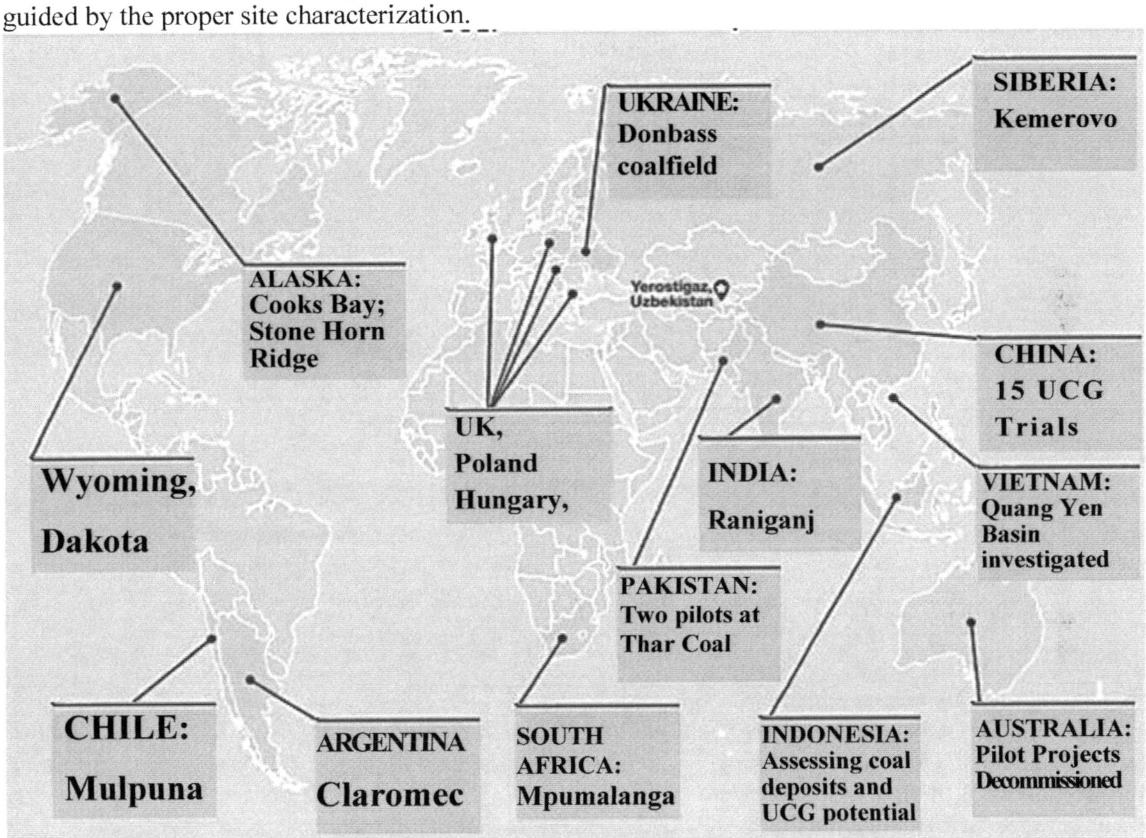

Figure 1. A brief snapshot of operational and prospective sites for UCG worldwide [6]

2. UCG Process

Underground Coal Gasification (UCG) is the process of in-situ gasification of coal aiming at recovery of heat value for power generation and, usable gaseous products as feed stocks for chemicals and fertilizers. The anticipated manifold increase in greenhouse gas emissions because of India's dependence on coal to meet the growing electricity demand may be mitigated economically by UCG combined with carbon capture and sequestration (CCS) than by addition of CCS to a coal-fired power plant. UCG also eliminates the hazards associated with conventional underground mining/surface strip mining or surface gasifiers operations. In contrast to conventional mining, UCG eliminates environmental issues of particular importance for utilization of low quality/high ash Indian coals as ash remains in the subsurface and needs no ash management, emits few black-carbon particulates, and it greatly reduces the need to mine and transport coal, since coal is used *in situ*. Overall, the C-footprint and capital expense associated with UCG much less compared to coal mining, transport, and surface gasification.

The 3 zones (Figure 2) of UCG gasifier from ignition point farther are:

Zone-I: COMBUSTION ZONE $C + O_2 \rightarrow CO_2$

Zone-II: REDUCTION ZONE $C + CO_2 \rightarrow CO$

Zone-III: PYROLISIS ZONE Coal $\rightarrow$ Volatiles $\rightarrow CO, H_2, CH_4$

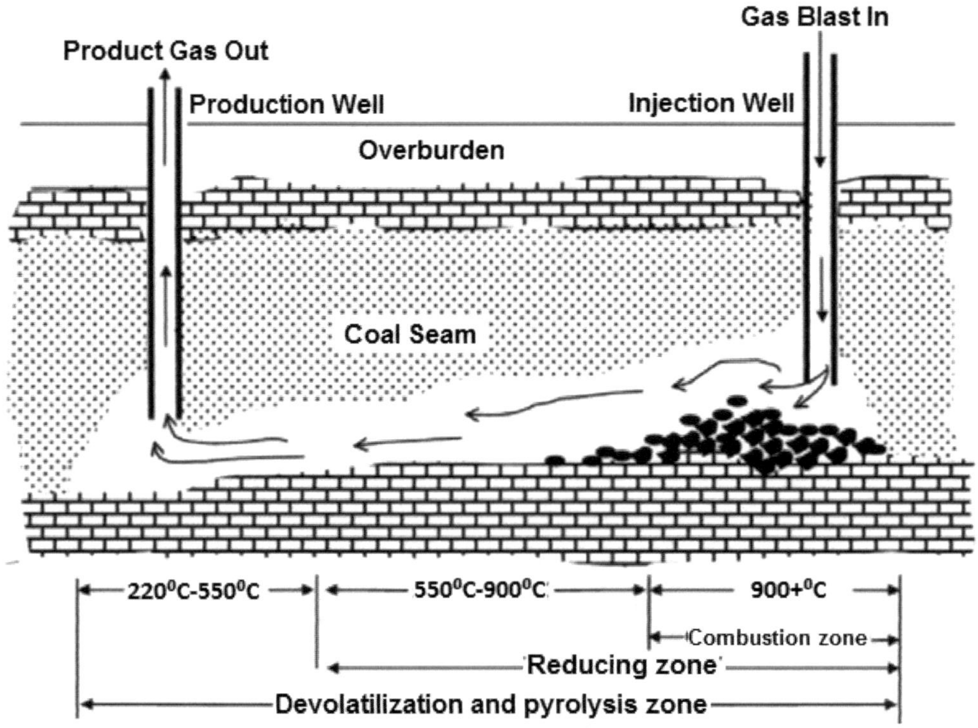

Figure 2. Schematic of in-situ gasification process [7]

3. Criteria for UCG site selection

UCG is a highly efficient coal conversion process [8], with gasification efficiencies similar to that of surface gasifiers (75-85%) and mining efficiencies comparable with that of underground coal mining (around 60%). Not all UCG trials were successful. Some early trials even caused geoenvironmental damage e.g. Hoe Creek trials of US. The key learnings from the trails are:

- Proper site selection
- Choice of UCG reactor design best suited for the site (*e.g.,* Figure 3)
- Operation in accordance to strict guidelines
- Monitoring and mitigation of associated geoenvironmental risks

UCG is very much similar to surface gasification as far as the process is concerned. However, UCG is limited by the fact that it is site specific, in terms of geology and geography of the area unlike surface gasifiers that can be designed, regulated, maintained and monitored in planned manner. Predicting the strata behaviour, monitoring cavity growth and, predicting and maintaining product gas quality are greatly limited by the poor control and monitoring of such complex process that varies over space and time. Hence, geology of the area plays decisive role throughout the life of a UCG project and imperative for every phase of the project cycle starting from planning, site selection, design to cessation of operations and restoration of the site [9]. Thus the criteria for UCG site selection are:

- Suitable coal seam characteristics for favourable operational conditions,

- Suitable geological and hydrogeological characteristics with negligible risk of groundwater pollution from the UCG operation, and
- Suitable geographic location that satisfies various surface installation requirements for operation.

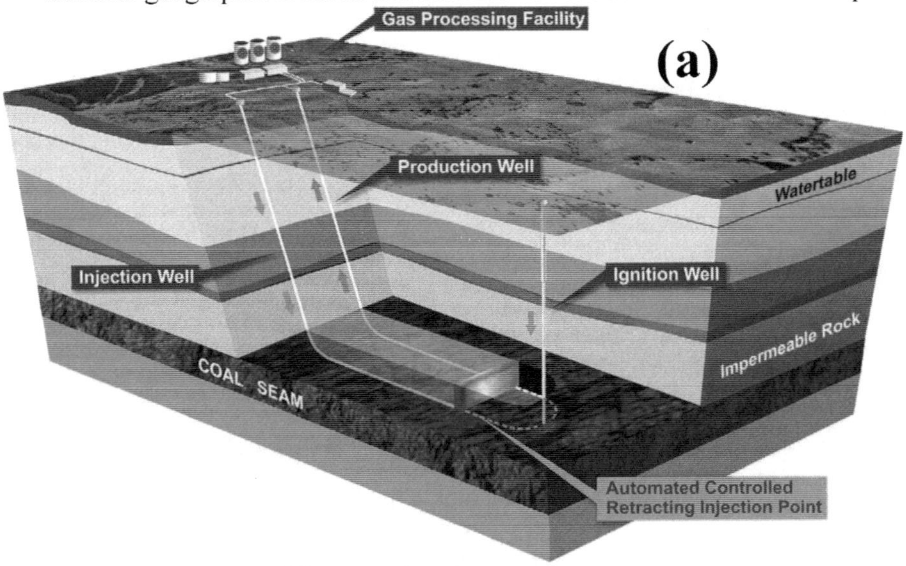

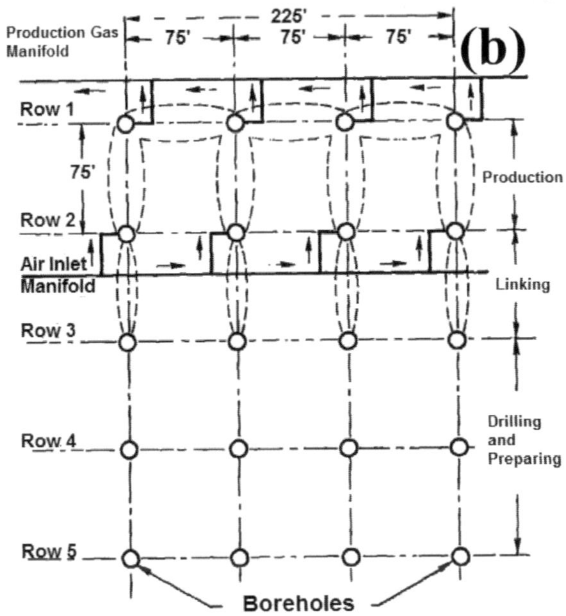

Figure 3: a. Carbon Energy's approach to UCG [10] b. plan view of a typical Soviet UCG process adapted for horizontal seams [11].

Understanding geology of the area provides necessary safeguards for protecting the environment. Geologically isolated deep beds are preferred to reduce the surface expressions of subsidence due to UCG cavity growth. Aquifers should be avoided in the vicinity of the UCG site. If at all this should be deep and consists of saline/non-potable water with stratigraphic seals. Structural integrity of overburden and coal should be insured to avoid possibility of gasifier roof caving in.

3.1. Coal seam geology

Coal deposits of different regions around the world differ remarkably in their geological setting, posing challenges for determination of selection criteria for a potential UCG sites. These factors include coal seam geology; physical, chemical and petrographic properties of coal; geomechanical properties of coal and surrounding rocks; and geo-hydrological regime of the area. Various geological (isopach/structure contour/isorank) maps, litholog, well log, seismic data etc. along with their spatio-temporal analysis and correlation (Figure 4) are useful for assessment and planning of UCG projects.

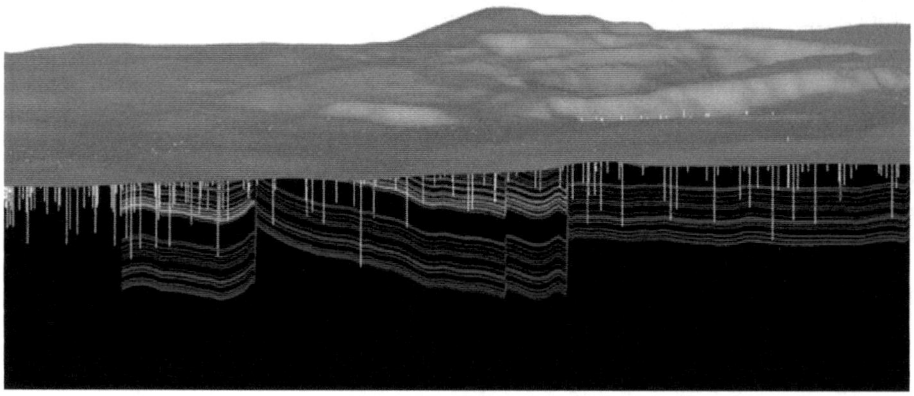

Figure 4: Showing subsurface geological modelling using various geological and geophysical data (from http://www.srk.com/)

Higher pressure in underground gasification chamber assures no groundwater inflow to the cavity. However, as pressure increases with depth of seam, deep seams with high coal and overburden permeability pose problem of enhanced product gas loss to the surrounding aquifer thereby altering the chemistry and contaminating the groundwater. An impermeable overburden helps provide a reasonable balance between pressure and gas losses.

3.2. Coal properties

Coal is a heterogeneous mixture of organic (macerals) and inorganic (minerals), and it is from the organic component that syngas is generated during gasification. The inorganic component has a significant impact on gasifier performance, where certain minerals can act as catalysts, others as inhibitors, depending on the gasification condition and type of mineral. Gasification of coal occurs in two stages, namely: (1) rapid pyrolysis devolatilizing the coal leading to char formation, and (2) char reaction. Reactivity in this context is the rate at which char will react with steam/oxidant to convert into syngas. Reactivity depends on chemical and physical (porosity) structure of char, and catalysts. Reactions occur only on the reactive surface area, and not all micropores may be reactive. If the heating rate is slow, devolatilization is followed by gasification; if the heating rate is fast, the two processes may occur simultaneously. Though all coal types of different ranks can be extracted and exploited through UCG process, low rank lignite and sub-bituminous are preferred for UCG being more permeable and reactive to gasification process. Some low coals shrink upon heating making the connection between the injections well and production well even greater. Higher ash content reduces gasifier efficiency due to heat loss and high O_2 demand. Leachability study of ash/slag is essential for disposal and environmental reasons.

Coal has dual porosity i.e., matrix/inherent (macro, meso, micro- pores) porosity and induced (cleat) porosity. The structural heterogeneity is the result of changes to the original organic matter due to varying PT conditions during the geological history. UCG depends on the porosity and permeability of the coal seam. Permeability is important in linking of injection and production well; better cleated and more permeable coal seams have greater ability to transport gases during gasification process while more porosity provide larger surface area for reactions to proceed effectively. Low rank coal are characterised by having porous and permeable structure.

3.3. Geological and hydrological conditions

The geological and geotechnical properties of surrounding strata as well as the coal seams have a direct bearing on the UCG process. Impermeable over/underlying strata with low porosity and less deformation are most suitable for UCG processes as they act as seal between the coal seam and the surrounding aquifers while limiting the degree of subsidence. Hydrological regime of the area forms an important screening criterion of potential UCG site. Water is needed for the UCG process as it takes part in the gasification reaction to add value to syngas composition. However inrush of excess water into the gasification chamber reduces the efficacy of the process and may even quench the reactions in progress. Presence of fresh water aquifer in the vicinity of target coal seam should be abandoned in order to avoid groundwater contamination. Risk of groundwater contamination arises mainly due to hot product gases escaping through crack and fissures while gasification operation is active. Contamination may also be caused due to post-burn leaching when gasification cavity fills with water. The case may be more devastating when the gasification chamber collapses thereby directly connecting coal to the aquifer. Hence coal seam suitable for a successful UCG process should preferably be in a vertical separation of more than 100m from the major aquifer.

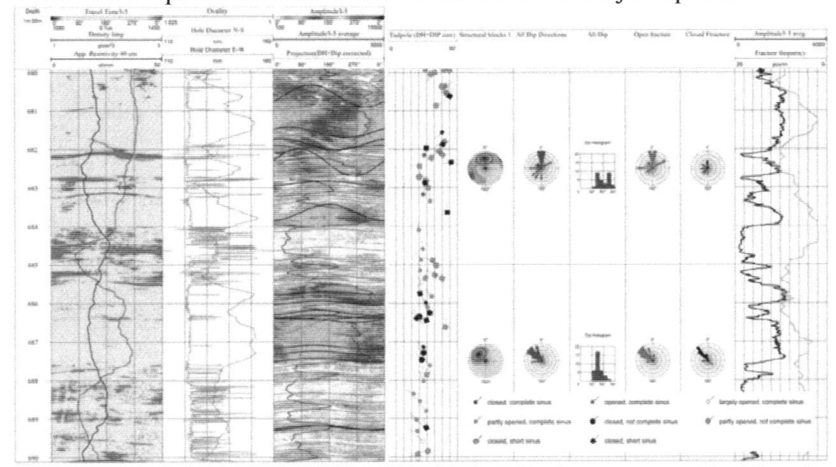

Figure 5. Acoustic Borehole Televiewer data showing borehole gauge, ovality, dip of drilling and its direction, attitude of fractures and fissures etc. (from [12])

4. Techniques used for UCG site investigation

UCG site selection requires proper assessment of depositional setting, tectonic/structural framework, hydrodynamics and coal type of the basin under consideration [13]. For this detailed geological information should be collected from various source. The detailed exploration programme includes literature information, outcrop mapping, borehole lithology, seismic survey, well log data etc. Subsurface geology is investigated with the help of borehole correlation. The borehole lithologs are supplemented with various downhole well log data. Coal seam geometry (mapping), petrophysical evaluation of coal and surrounding rock, geomechanical parameters, cleat/fracture system characterisation, structural analysis etc. are some of the prime objective of down-hole geophysical investigations. Various well logging techniques deployed now-a-days are outlined below:

- Resitivity/SP logs: To demonstrate occurrence of sand with water and clay bands based on formation resistivity
- Gamma-Gamma log: To evaluate formation density
- Gamma Ray log: To detect clay/mudrock layers using natural radioactivity
- Neutron log: To determine porosity
- BHTV (Borehole Televiewer): To locate fractures and determine hole condition (Figure 5)
- OPTV (optical imagery): To monitor borehole walls

Table 1. Some of the instrumental techniques used for coal/rock characterization

Generic Name	Usage
CHNS analyzer	Chemical Characterization
Petrological microscope	Maceral/Reflectance

Generic Name	Usage
Proximate Analyser	Chemical Characterization
BET surface area analyzer	Surface characterization
Ultra pycnometer – 1000	Porosity study
Mercury porosimete	Porosity study
CPT/IPT Apparatus	Crossing Point and Ignition Point measurements
XRD	Mineralogical Study
FE-SEM EDAX	Morphological/Mineralogical Study
Uniaxial compressive strength Testing Facility	Physico-Mechanical Study
Triaxial strength Testing Facility	Physico-Mechanical Study

Locating faults, defining lateral extent and nature of strata at the site needs critical desk study of the available data that may include *in situ* hydrogeological data and detailed seismic data along with borehole litholog and well log data. On site hydrological investigation includes [14] borehole water pressure and its deviation from hydrostatic head, transmisivity and water quality (mainly total dissolved solids/conductivity). Different instrumental techniques used for coal/rock characterization are as given in Table 1.

5. Monitoring UCG process

UCG monitoring was initially carried out during the 1930's in the former Soviet Union (FSU) with thermocouples, flow rate meters and gas analysis equipment. Efficient and safe operation of UCG requires strategic UCG monitoring program which includes the followings:

- Process parameters such as gas flow rates both for oxidant injection and syngas production, temperature, pressure and syngas content;
- Geomechanical parameters such as alteration in stress regime through coal seam, overburden and cavity pressures, cavity development, subsidence and ground deformation etc.; and
- Environmental parameters such as change in water table, groundwater chemistry and ambient air quality etc.

The strategic UCG monitoring program should be done in three stages:

- *Assessment and Planning*: During this phase, the site is characterized geographically and geologically. Based on these studies, an array can be designed to meet the requirements of regulators and other stakeholders.
- *Baseline Monitoring*: Reliable pre-burn baseline surveys must be carried out before hand to understand the background values and to provide a basis for difference mapping.
- *Operational Monitoring*: Regular and frequent sampling during the burn for near real-time monitoring of injection/production wells to look for circulation behind casing, failures within the well bore and, to observe ground deformation, ground water contamination and other operational problems or failures.
- *Array Monitoring During and After Gasification*: Long-term post-burn monitoring phase involves active surface and subsurface arrays, with the option for additional tools around high-risk zones. The recurrence and total duration of monitoring is to be determined by the research goals, the site parameters, the budgetary status and regulatory needs.

Monitoring program requires appropriate sensors for comprehensive spatial as well as temporal coverage of an entire UCG process. Lateral and vertical network of near, middle and far monitoring wells are required to collect data and samples from the coal seam, overburden and underburden. Apart from routine monitoring instrumentation, modern technologies are available those can be easily be deployed to generate high density 2D/3D data e.g., electromagnetic imaging, resistance tomography, seismic and magnetotelluric techniques can be used for detection of burn front and delineation of cavity growth [15]. Extensometers, shear strips and piezometers are generally used to monitor surface deformation. New techniques such as tilt meter and InSAR are now-a-days available for monitoring deformation and subsidence, and fiber optic temperature sensors for monitoring cavity temperature.

Cavity growth due to *in situ* gasification often leads to roof collapse leading to subsidence and establishing connection between the UCG cavities with overlying aquifers. An active groundwater monitoring program is required to address the concerns of groundwater contamination during UCG operation [16]. Borehole hydrological monitoring, geochemical and tracer techniques, geophysical monitoring (seismic, electric, magnetic and gravity) methods should be variously deployed for monitoring the operation, shut down and restoration. Latest in the sequence the ambient air quality monitoring for UCG is required to identify potential pollutants escaped from the gasifier.

6. Shutting down

Appropriate shutdown procedure is highly imperative to avoid issues of possible contamination as pyrolysis products will no longer be consumed on cessation of reaction following non-injection of oxidants. Effective shutdown programme should intend at minimising the post-burn contaminant generation from pyrolysis products by accelerating the cooling of the cavities and preventing post gasification pressure build up. This may be achieved by pumping out water from the cavities to maintain hydrostatic gradient towards the reactor to flush out potential organic and inorganic contaminants from the underground strata and treating contaminants, and to restore ground water quality to near baseline conditions.

References

[1] IEA 2014 *Medium-Term Coal Market Report 2014: Market Analysis and Forecasts to 2019* (Paris: International Energy Agency)
[2] Friedmann S J, Upadhye R and Kong F M 2009 *Energy Procedia* **1** 4551–4557
[3] Stanczyk K, Howaniec N, Smolinski A, Świądrowski J, Kapusta K, Wiatowski M, Grabowski J, and Rogut J 2011 *Fuel* **90** 1953–1962
[4] Bhutto A W, Bazmi A A and Zahedi G 2013 *Progress in Energy and Combustion Science* **39** 189–214
[5] Younger P L 2011 *Mine Water and the Environment* **30** 127–140
[6] Yang D, Sheng Y and Green M 2014 *The Chemical Engineer* (872) 38-41
[7] Lee S 2015 *Handbook of Alternative Fuel Technologies – Second Edition* ed S Lee, J Speight and S K Loyalka (Boca Raton: CRC Press) p 19-84
[8] Cena R J, Britten J A and Thorness C B 1988 *14th UCG Annual Symposium* (Chicago)
[9] Mohanty D 2004 *Cognizance-2004* (IIT Roorkee)
[10] Morné E 2015 *Corner Stone, The Official Journal of the World Coal Industry* **3**(2) 61-64
[11] Gregg D W, Hill R W and Olness DU 1976. *An Overview of the Soviet effort in underground coal gasification* (Livermore: LLNL) Report No. UCRL-52004.
[12] Burján B and Mázik J 2011 *6th Congress of the Balkan Geophysical Society* (Budapest)
[13] Burton E, Friedmann J and Upadhye R 2004 *Best Practices in Underground Coal Gasification* (Livermore: LLNL) Contract No. W-7405-Eng-48
[14] Sury M, White M, Kirton J, Peter C, Woodbridge R, Mostade M, Chappell R, Hunt D, Rendell N and Hartwell D 2004 *Review of Environmental Issues of Underground Coal Gasification – Best Practice Guide* () Report No. COAL R273 DTI/Pub URN 04/1881
[15] Yang X, Wagoner J and Ramirez A 2012 *Monitoring of Underground Coal Gasification* (Livermore: LLNL) Report No. LLNL-TR-578172
[16] Creedy D P, Garner K, Holloway S, Jones N and Ren T X 2001 *Review of underground coal gasification technological advancements*. (London: UK DTI Cleaner Coal Technology Transfer Program) Report No. COAL R211, DTI/Pub URN 01/1041

Acknowledgement

The author is grateful to CSIR Network Project CoalGasUrja (ESC-0302) for financial support. He also thanks Dr. Pradeep K Singh, Director, CSIR-CIMFR, for his support and guidance in preparation of this paper.

IOP Conf. Series: Earth and Environmental Science **76** (2017) 012011 doi:10.1088/1755-1315/76/1/012011

CO_2 Capture and Storage in Coal Gasification Projects

Anand B Rao[1,2,3,*] and Pranav C Phadke[1]

[1]Centre for Technology Alternatives for Rural Areas (CTARA), Indian Institute of Technology Bombay, Powai, Mumbai 400076, INDIA
[2]Department of Energy Science & Engineering (DESE), Indian Institute of Technology Bombay, Powai, Mumbai 400076, INDIA
[3]Inter-disciplinary Program in Climate Studies, Indian Institute of Technology Bombay, Powai, Mumbai 400076, INDIA

*Corresponding Author. Tel: +91-22-25767877, Fax: +91-22-25767874,
E-mail: a.b.rao@iitb.ac.in

Abstract. In response to the global climate change problem, the world community today is in search for an effective means of carbon mitigation. India is a major developing economy and the economic growth is driven by ever-increasing consumption of energy. Coal is the only fossil fuel that is available in abundance in India and contributes to the major share of the total primary energy supply (TPES) in the country. Owing to the large unmet demand for affordable energy, primarily driven by the need for infrastructure development and increasing incomes and aspirations of people, as well as the energy security concerns, India is expected to have continued dependence on coal. Coal is not only the backbone of the electric power generation, but many major industries like cement, iron and steel, bricks, fertilizers also consume large quantities of coal. India has very low carbon emissions (~ 1.5 tCO_2 per capita) as compared to the world average (4.7 tCO_2 per capita) and the developed world (11.2 tCO_2 per capita). Although the aggregate emissions of the country are increasing with the rising population and fossil energy use, India has a very little contribution to the historical GHG accumulation in the atmosphere linked to the climate change problem. However, a large fraction of the Indian society is vulnerable to the impacts of climate change – due to its geographical location, large dependence on monsoon-based agriculture and limited technical, financial and institutional capacity. Today, India holds a large potential to offer cost-effective carbon mitigation to tackle the climate change problem. Carbon Capture and Storage (CCS) is the process of extraction of Carbon Dioxide (CO_2) from industrial and energy related sources, transport to storage locations and long-term isolation from the atmosphere. It is a technology that has been developed in recent times and is considered as a bridging technology as we move towards carbon-neutral energy sources in response to the growing concerns about climate change problem. Carbon Capture and Storage (CCS) is being considered as a promising carbon mitigation technology, especially for large point sources such as coal power plants. Gasification of coal helps in better utilization of this resource offering multiple advantages such as pollution prevention, product flexibility (syngas and hydrogen) and higher efficiency (combined cycle). It also enables the capture of CO_2 prior to the combustion, from the fuel gas mixture, at relatively lesser cost as compared to the post-combustion CO_2 capture. CCS in gasification projects is considered as a promising technology for cost-effective carbon mitigation. Although many projects (power and non-power) have been announced internationally, very few large-scale projects have actually come up. This paper looks at the various aspects of CCS applications in gasification projects, including the technical feasibility and economic viability and discusses an Indian perspective. Impacts of including CCS in gasification projects (e.g. IGCC plants) have been assessed using a simulation tool. Integrated Environmental Control Model (IECM) – a modelling framework to simulate power plants - has been used to estimate the implications of adding CCS units in IGCC plants, on their performance and costs.

1. Introduction

Close to one third of the population in India has no access to reliable electricity. The per capita annual energy consumption in India is ~ 1075 kWh [1] whereas the global average is ~ 3200 kWh [2]. The required electricity is presently being supplied by coal, natural gas, hydro, nuclear power, and renewable energy, with coal contributing to about 60% [1]. Table 1 provides a snapshot about current installed capacity in the country as of 31st Dec 2016.

Table 1. Total installed capacity in India as of 31st Dec 2016 [1]

Energy Source	Installed Capacity (MW)	% of Total
Coal	188,968	60.0%
Gas	25,282	8.2%
Oil	919	0.3%
Hydro	43,139	13.9%
Nuclear	5,780	1.9%
Renewable Energy Sources*	45,917	14.8%
Total	**310,005**	**100%**

*Renewable Energy Sources include small hydro project, biomass, waste to electricity, wind energy, solar (PV + thermal)

As per the NITI Ayog report [3], coal is expected to continue to be a major energy source in India for the next few decades due to the large demand of energy requirement of the economy, and adequate availability of coal. India has substantial reserves of low sulfur, low grade (high ash) coal, which provides a reliable, cheap baseload power. But the coal-based thermal power industry is responsible for significant share of emissions of the industrial sector in India. Main emissions from coal fired thermal power plants are CO_2, NO_x, SO_2, and fly ash [4]. Between 2001-02 to 2009-10, emissions from coal power plants have increased by 4.9% for CO_2, 4.7% for SO_2 and 4.9% for NO_x measured in CAGR (compounded annual growth rate) [4]. Recognizing these worrying trends, several countries have imposed emission control norms for coal power plants. In India, the Ministry of Environment, Forest & Climate Change (MoEF&CC), in December 2015, announced more stringent emission control standards for coal-based thermal power plants in order to drastically reduce emissions of particulate matter (PM), sulfur dioxide (SO_2), oxides of nitrogen (NO_x). However, CO_2 emissions from coal power plants in India are currently unregulated. Coal power plants have the highest density of CO_2 emissions when normalized per unit area and unit time [5]. Recognizing that CO_2, a major greenhouse gas responsible for climate change, countries like Canada, US, UK have already imposed CO_2 emission standards for coal power plants [6]. Thus, for climate change mitigation while continuing the use of fossil fuels, a special attention needs to be given to the reduction of CO_2 emissions by capture and storage (CCS) techniques applied to power generation and other energy intensive industrial sectors (cement, metallurgy, oil refinery etc.) [7].

Carbon capture and storage or carbon capture and sequestration (CCS) involves capture and separation of CO_2 as a concentrated stream for subsequent transport and storage. Thus, the concept of carbon capture and sequestration (CCS) can be divided into three major steps namely: CO_2 capture, transport and storage. The separation stage (i.e. CO_2 capture) is the most energy intensive and largely determines the cost of the CCS as a whole. It represents ~ 75–80% of the total cost of CCS [8]. The three potential technology options which have a potential to handle CO_2 emissions from large-scale point sources, such as the fossil fuel based power plants are: a) Post-combustion capture of CO_2 from the flue gas, b) Pre-combustion capture of CO_2 from the fuel gas, and c) Oxyfuel combustion for CO_2 capture [9]. Subsequently, the captured CO_2 is then transported using pipelines, ships, trucks and trains to suitable storage sites. There are broadly three ways of sequestering CO_2: geological storage, ocean and mineral fixation.

Most available CCS technologies capture about 85–95% of the CO_2 processed in a capture plant. A power plant equipped with a CCS system (with access to storage) would need roughly 10–40% more energy than a plant of equivalent output without CCS, of which most is for capture and compression. With secure storage, the net result is that a power plant with CCS could reduce CO_2 emissions from power plant to the atmosphere by approximately 80–90% compared to a plant without CCS [8]. Thus, CCS can be a key tool in tackling climate change and would become an unavoidable option if we are to ensure that we can meet the increasing electricity demand with an acceptable carbon footprint.

Post-combustion capture of CO_2 from the flue gas is employed in case of conventional power plants. However, before implementation of CCS, these power plants also require other emission control technologies for stringent SO_2 and NO_x control. This further adds to the cost of electricity.

An alternative technology to generate power from coal is through the gasification process. Here, the coal is converted into a gaseous mixture (syngas) in a gasifier, which is then subjected to a gas turbine to generate electricity. Additional power is generated from a steam turbine using steam generated from the het recovered from the flue gases of the gas turbine. It is possible to undertake pre-combustion capture of CO_2 from this "Integrated Gasification Combined Cycle (IGCC) plant" where CO_2 can be separated from syngas at a much lower energy penalty, also resulting in much lower emissions of conventional pollutants as compared to the conventional power plants. In this paper, we focus on this advanced coal technology and the feasibility of pre-combustion capture of CO_2 in terms of performance, emissions, capital cost and levilised cost of electricity (LCOE).

2. CCS in Coal Gasification Projects

Integrated Gasification Combined Cycle (IGCC) is emerging as a promising technology to utilize low quality or contaminated energy resources such as coal. It is a technology that uses a high-pressure gasifier to turn coal and other carbon based fuels into pressurized gas—synthesis gas (syngas) and can then remove impurities from the syngas prior to the power generation cycle. This results in lower emissions of SO_2, particulates, mercury, and even CO_2. Thus, by integrating CCS with IGCC plants, drastic CO_2 emission reductions (81-91%) can be achieved [8]. The schematic of a typical IGCC plant with CCS has been shown in Figure 1.

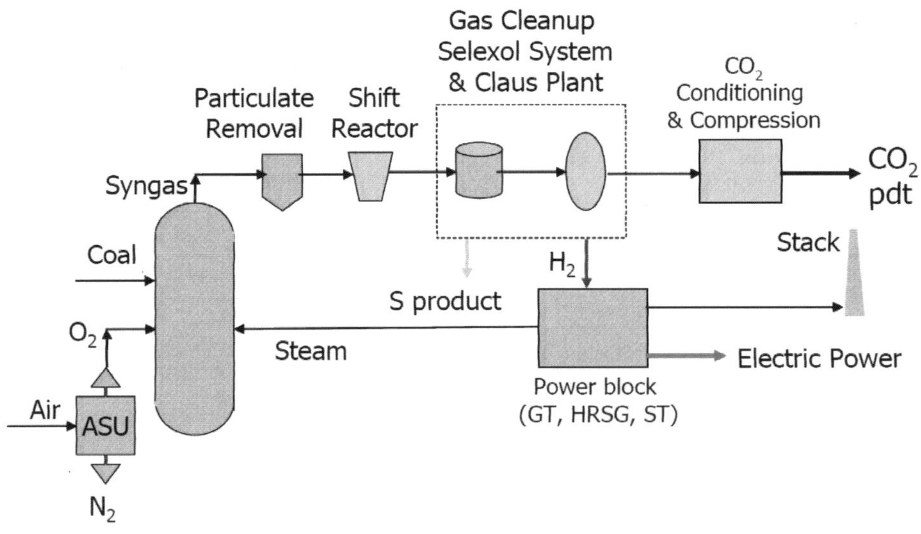

Figure 1. Schematic of IGCC Plant with CCS [9]

IOP Conf. Series: Earth and Environmental Science **76** (2017) 012011 doi:10.1088/1755-1315/76/1/012011

Initially, coal is fed into a high-pressure vessel called gasifier with controlled quantity of oxygen (or air) and steam (or water) to produce hot syngas. This hot syngas is then cooled and filtered to remove particulate matter. The heat extracted is used to generate high-pressure steam. The sour water gas shift (WGS) reaction is a key conversion step in the IGCC-CCS process in which carbon monoxide (CO) produced by the gasification of the fuel is reacted with steam to produce additional hydrogen and CO_2 in the presence of H_2S [9, 10]. The resulting acid gas stream consisting of H_2S, CO_2 and H_2O is sent to the Selexol™ system and Claus unit where H_2S is recovered as elemental sulfur, which is a valuable byproduct. Most of the CO_2 in the flue gas stream is then recovered, compressed, transported and sequestered.

2.1 Advantages of IGCC with CCS

1. Less pollution: Significant reduction in emission of criteria air pollutants (SO_2 and NO_x) compared to Pulverized Coal (PC) boilers.
2. Lower energy penalty: Typically, the syngas is obtained at high pressure with substantial CO_2 content and hence, less energy required for the separation and capture of CO_2 from the syngas.
3. Multi-product flexibility: Hydrogen (H_2) can be consumed in the gas turbine to generate power, or as a valuable chemical agent with a great variety of industrial applications.
4. Valuable byproducts: Sulfur (almost all the sulfur in the coal can be recovered), Nitrogen (from the air separation unit) and CO_2 (from the CO_2 capture unit) are also obtained as byproducts [9].

2.2 Disadvantages of IGCC with CCS:

1. Gasification is a complicated chemical process and not many IGCC plants exist around the globe. For a given coal, the resulting product (syngas) depends on fine-tuning of a variety of control parameters including the gasifier design, the gasifier temperature and pressure, and the amount of steam and air/ oxygen input.
2. Uncertainty about the capture ready status of IGCC plants with CCS for Indian scenarios. Pilot projects in Indian context required. Also, compared to post-combustion capture, it is not easy to retrofit an existing IGCC plant to accommodate CO_2 capture unit, since it will affect the power generation from the turbines.
3. Uncertainty about use of low quality coals for IGCC plants with CCS.
4. Need for a large-scale demonstration project to establish the commercial viability of the technology
5. Higher capital cost of IGCC plants compared to PC plants, lack of experience of CCS, lack of accurate estimation of storage potential and the perceived financial risk [9].

3. Current Global Status of Gasification based CO_2 Capture

Table 2 lists some of the gasification projects that are operational/under construction/under evaluation globally which employ pre-combustion (gasification) CO_2 capture & storage.

Table 2. Some of the gasification projects that are operational/under construction/under evaluation globally which employ Pre-combustion (gasification) CO_2 capture & storage [11]

Project name	Location	Purpose	CO_2 Capture Start	Capture capacity (Mtpa)	Transport type	Primary storage type
Great Plains Synfuel Plant and Weyburn-Midale Project	Canada	Production of natural gas and other by-products	2000	3.0	Pipeline	Enhanced oil recovery

Kemper County Energy Facility	Mississippi, United States	Power Generation	2017	3.0	Pipeline	Enhanced oil recovery
Yanchang Integrated Carbon Capture and Storage Demonstration Project	China	Chemical Production	2018	0.4	Pipeline	Enhanced oil recovery
Texas Clean Energy Project	Texas, United States	Power Generation	2021 (estimate)	2.4	Pipeline	Enhanced oil recovery
Huaneng GreenGen IGCC Project (Phase 3)	China	Power Generation	2020's (planned)	2.0	Pipeline	Enhanced oil recovery
Caledonia Clean Energy Project	United Kingdom	Power Generation	2022 (planned)	3.8	Pipeline	Dedicated Geological Storage
Coffeyville Gasification Plant	Kansas, United States	Fertiliser production	2013	1	Pipeline	Enhanced oil recovery

Out of these projects, Kemper county energy facility is the only large scale IGCC+CCS project for electric power generation, which is expected to start CO_2 capture soon. A 582 MW plant using TRIG™ Integrated Gasification Combined Cycle is set to capture at least 65% of CO_2 emissions and use it for enhanced oil recovery by sequestering it in Mississippi oil fields, Mississippi. The total cost of the project, designed to generate electricity from burning 4.5 million tonnes of Mississippi lignite coal per year, is around $7.1 billion [11, 12].

4. Plant Simulations

The impact of integrating CCS in an IGCC plant has been demonstrated here through a case study simulation. Using Integrated Environmental Control Model (IECM) – a modelling framework to simulate power plants [13], we simulated an IGCC plant with and without CCS. Plants are simulated using both bituminous and sub-bituminous coal, whose composition is provided in Table 3.

Table 3. Characteristics of Coals considered for the Case Study

Bituminous Coal									
Moisture		Ash		C	H	N	S	O	HHV (MJ/kg)
11.12		9.70		63.75	4.50	1.25	2.51	6.88	27.14
Ash Analysis (wt. %)									
SiO_2	Al_2O_3	Fe_2O_3	P_2O_5	TiO_2	CaO	K_2O	Na_2O	MgO	SO_3
46.80	18.00	20.00	0.20	1.00	7.00	1.90	0.60	1.00	3.50
Sub-Bituminous Coal									
Moisture		Ash		C	H	N	S	O	HHV (MJ/kg)
30.24		5.32		48.18	3.31	0.70	0.37	11.87	19.40
Ash Analysis (wt. %)									
SiO_2	Al_2O_3	Fe_2O_3	P_2O_5	TiO_2	CaO	K_2O	Na_2O	MgO	SO_3
63.19	30.00	2.90	0.08	0.09	0.91	1.49	0.38	0.76	0.20

The details about the plant configurations and some key input parameters used in IECM simulations are listed in Table 4.

Table 4. Configuration of IGCC plant with and without CCS

Parameters	**IGCC**	**IGCC + CCS**	**IGCC**	**IGCC + CCS**
Gross Capacity (MW)	724.3	690.3	748.5	691.4
Coal	Bituminous		Sub-Bituminous	
Capacity Factor (%)	75			
Gasifier Temperature (°C)	1343			
Gasifier Pressure (Mpa)	4.24			
Turbine Inlet Temperature (°C)	1371			
Cost of Fuel ($/tonne)	42.09		9.64	
Annual operating hours (hours)	6575			
H_2S Control	Selexol^{TM}			
H_2S Removal Efficiency (%)	98			
Plant Life (years)	30			
CO_2 Capture	-	Sour Shift + Selexol^{TM}	-	Sour Shift + Selexol^{TM}

The composition of the syngas obtained from the gasifier after the gasification of coal and the performance and cost results have been provided in Table 5 and Table 6, respectively.

Table 5. Syngas composition obtained from gasifier

Syngas composition obtained using bituminous coal							
CO	H_2	CH_4	H_2S	CO_2	H_2O	N_2	Ar
30.64	32.92	0.26	0.97	18.52	14.86	0.86	0.87
Syngas composition obtained using sub-bituminous coal							
CO	H_2	CH_4	H_2S	CO_2	H_2O	N_2	Ar
22.66	31.06	0.10	0.12	22.01	22.53	0.69	0.79

Table 6. Simulation Results for IGCC plant with and without CCS

Parameters	**IGCC**	**IGCC + CCS**	**IGCC**	**IGCC + CCS**
Net Capacity (MW)	609.0	529.7	620.0	513.3
Gross Capacity (MW)	724.3	690.3	748.5	691.4
Coal Type	Bituminous		Sub-Bituminous	
Cost of Fuel ($/tonne)	42.09		9.64	
Cost of Fuel ($/MJ)	1551		497.20	
Gross Plant Heat Rate, (kJ/kWh)	9247	10130	9448	10370
Net Plant Heat Rate, (kJ/kWh)	10990	13200	11410	13970
Net Plant efficiency (%)	32.73	27.27	31.56	25.78
Coal Flow Rate (kg/s)	68.55	71.58	101.27	102.66
Total Makeup Water Requirement (m^3/MWh)	1.55	1.76	1.55	1.81
Emissions				
CO_2 emission (kg/MWh_{net})	918.39	93.10	1006.93	94.07
SO_2 emission (kg/MWh_{net})	0.57	0.07	0.09	0.01
NO_x emission (kg/MWh_{net})	0.05	0.06	0.05	0.06
Particulate emission (kg/MWh_{net})	0.004	0.005	0.004	0.006
Cost (in 2014 USD)				
Total Capital Cost (M$)	1643	2177	1709	2259
Fixed O&M M$/year	52.90	73.08	55.26	76.83
Variable O&M M$/year	74.91	94.09	32.21	49.50
LCOE ($/MWh)	78.21	118.5	68.75	112.9
CO_2 Avoidance Cost ($/tonne)	-	49.30	-	48.73

IOP Conf. Series: Earth and Environmental Science **76** (2017) 012011 doi:10.1088/1755-1315/76/1/012011

It can be seen that it is difficult to keep the same power output of the IGCC plant, if CCS is incorporated. Using CCS, the CO_2 emissions from the IGCC plants can be reduced by about 90% as compared to that of the IGCC plant without CCS. However, the levelised cost of electricity (LCOE) increases by about 51% for the bituminous coal case and to about 64% for the sub-bituminous coal case. Thus, CCS can substantially reduce the CO_2 emissions though it is significantly expensive at present.

5. Barriers and Issues Related to CCS

CCS is still in developmental stages globally and following are some of the main barriers in adoption of this technology [14]:

1. Technical Issues: Worldwide still there aren't many large scale successful CCS projects and thus on-field experience is still lacking with this technology. There is a lack of accurate geological storage site data and depending upon the location and plant size, feasibility studies are required. Also, there is no clarity about the feasibility of CCS retrofits for the existing plants, especially in case of IGCC plants.
2. Economic Issues: CCS technology drastically increases the levelised cost of electricity and also decreases the net plant output. Enhanced Oil Recovery (EOR) is currently the only attractive option for CO_2 storage, since the cost of storing the CO_2 is (partly) offset by the revenues accrued by oil recovered from the depleted/depleting oil fields. Access to funding from financing agencies and incentives are required.
3. Social Issues: Public acceptability of CCS is not yet understood.
4. Monitoring Issues: There is a risk of slow/accidental leakages of CO_2 to atmosphere. This poses a further environmental, health and safety risk. Rigorous monitoring is needed over long period of time.
5. Legal Issues: Issues related to land acquisition, ground water contamination, CO_2 leakages and long term liabilities need to be resolved.
6. Regulatory and Strategic Issues: Specific clearances would be required in addition to existing clearances and a smooth regulatory framework would be required.

6. Conclusion

CCS is a potential mitigation option to tackle global climate change problem, but it needs to overcome several barriers/challenges. The key challenges are technical feasibility, assessment of storage potential, scale up of technology, economic viability, public acceptance and monitoring issues. Pilot plants as well as demonstration projects are now under development across many countries. IGCC plant with CCS is an attractive option as it also reduces SO_2 and NO_x emissions significantly and imposes (relatively) lower energy penalty for separation and capture of CO_2, as compared to combustion plants. However, the technology is yet to be demonstrated at the full scale. Gasification using the high ash containing coals in India is a challenge [15]. The issue of CCS drastically increasing the cost of electricity while reducing net power output poses the biggest barrier to acceptability of CCS in India [13]. Our priorities as a developing country - towards improving efficiency, reducing costs and increasing electricity accessibility - are in odds with that of the expected impacts of CCS implementation. However, R&D in CCS need to continue in search for more acceptable solution so as prepare ourselves for the future.

References

[1] Central Electrical Authority 2017 *Executive Summary* available at http://www.cea.nic.in/reports/monthly/executivesummary/2017/exe_summary-01.pdf

[2] The World Bank n.d., *Electric Power Consumption (kWh per capita)*, available at http://data.worldbank.org/indicator/EG.USE.ELEC.KH.PC

[3] NITI Aayog 2015 *A Report on Energy Efficiency and Energy Mix in the Indian Energy System (2030), Using India Energy Security Scenarios, 2047* http://niti.gov.in/content/report-energy-

efficiency-and-energy-mix-indian-energy-system-2030-using-india-energy
[4] Mittal M L, Sharma C amd Singh R 2014 *Environmental monitoring and assessment* **186** 6857-6866
[5] White C M, Strazisar B R, Granite E J, Hoffman J S and Pennline H W 2003 *Journal of the Air & Waste Management Association* **53** 645-715
[6] Electric Power Research Institute 2015 *Can Future Coal Power Plants Meet CO_2 Emission Standards Without Carbon Capture & Storage?* http://www.epri.com/abstracts/Pages/ProductAbstract.aspx?ProductId=000000003002006770
[7] Cormos C C 2012 *Energy* **42** 434-445
[8] IPCC 2005 *IPCC special report on carbon dioxide capture and storage* (Cambridge University Press)
[9] Rao A B 2007 *Carbon Capture and Sequestration: Integrating Technology, Monitoring, and Regulation* ed E J Wilson and D Gerard (Iowa: Blackwell Publishing) p 11-34
[10] Beavis R, Forsyth J, Roberts, E, Song B, Combes G, Abbott J and Barton I 2013 *Energy Procedia* **37** 2256-2264
[11] Global CCS Institute *Projects Database* https://www.globalccsinstitute.com/projects/large-scale-ccs-projects#map
[12] Misissipi Power *Kemper County Energy Facility, Quick Facts* http://www.mississippipower.com/about-energy/plants/kemper-county-energy-facility/facts
[13] Integrated Environmental Control Model (IECM) *Version 9.2.1* (Carnegie Mellon University) https://www.cmu.edu/epp/iecm/
[14] TERI *Barriers to CCS implementation in India* available at https://hub.globalccsinstitute.com/publications/india-ccs-scoping-study-final-report/8-barriers-ccs-implementation-india
[15] Singh N, Raghavan V and Sundararajan T 2013 In *9th Asia-Pacific Conf. on Combustion, Korean Combustion Society* (*Gyeongju, Korea*)

Acknowledgements
The authors are grateful to the GTWG-ACT project of the Department of Science and Technology (DST), Government of India, for financial support. Authors also thank the IECM team at the Carnegie Mellon University, USA, for making the tool freely available.

IOP Conf. Series: Earth and Environmental Science **76** (2017) 012012 doi:10.1088/1755-1315/76/1/012012

Policy and Regulatory Issues for Underground Coal Gasification in India

Sunil K Singh

Alternate Energy Department, Directorate General of Hydrocarbons, Noida, India

* Corresponding Author. Email: ssingh@dghindia.gov.in

Abstract. Underground coal gasification (UCG) is in its nascent stage of development. Most of the projects are in the nature of pilot projects. UCG technology requires acceptance in general commercial framework as it matures with the progress of time. Policy and regulatory framework, therefore, is considered here only in the expectation that UCG technology may finally be rolled out sooner than later. India is actively pursuing consultations with major countries which have recorded successes in implementing UCG technology in varying measures. In this background, the discussion on policy and regulatory framework is essentially an effort to capture the broad outline of the understanding of the UCG process in a regulatory construct as compared with other regulatory regimes of similar nature.

1. Introduction

Underground coal gasification (UCG) is being developed internationally as a leading source of unconventional gas reserve. Number of initiatives internationally are focussing upon the creation of framework for UCG for its safe, environmental-friendly and economic deployment. Thus, a background of norms, rules and safety practices are required to bring UCG to the forefront of energy security in India. Many reasons can be cited for the delayed development of UCG in India. However, a robust background of policy framework can be said to be a major reason attributable to slower-than-usual development of gasification based economy in India. Both underground coal gasification and surface gasification have not reached a commercially viable level despite various efforts by the governmental, policy, industrial and academic stakeholders. It should be noted that while early efforts for UCG were taken up in the early 1980's itself with concerted inputs of Coal India and ONGC. One lignite block in Rajasthan was found to be suitable but pilot operations could not begin due to fears of impending ground water contamination [1]. Similarly, surface coal gasification efforts began in the erstwhile CFRI in the 1960's [2]. Iyengar and Haque [3] and Krishnudu et al [4] performed trials on laboratory scale for understanding of gasification processes in Indian high-ash coals. However, as Shahi [5] points out, gasification never really "took-off" so as to meet societal needs in sustainable manner. It is important to understand that proper implementation internationally has been accompanied by proper regulatory framework. In India itself, development of coalbed methane (CBM) and mine drainage has been successful with sufficient policy inputs. Thus, understanding the modalities in terms of UCG feasibility in terms of rules and acts, environmental norms and so on need to be understood. This paper intends to understand the global policy developments in UCG technology through a overview of some of the work performed in this area. It should be noted that this is only a preliminary understanding of the work ahead. For instance, the Government of India recently issue new emission norms for coal-fired power plants. Should UCG-based power plants be subject to similar or lower norms in terms of particulate and gaseous emissions? If UCG is expected to be accompanied with CO_2 capture and storage (CCS), would the regulations for UCG-CCS be similar to PC-CCS plants and IGCC-CCS plants? If not, what are the possible areas where regulations may be different? Further, since conventional coal mining activities led to significant fugitive methane emissions [5], the Directorate General of Mines Safety (DGMS) has established norms for degrees of gassiness of coal mines. Similar classifications also need to be developed for UCG projects.

2. Summary of international policy studies

Mao [6] reviewed the important factors related to UCG development in China. Some of the important highlights of the study are shown in Table 1.

Table 1. Highlights of Mao [6] in terms of UCG prospects

Area	Highlights
Environment and safety	• Approval of over eighty eco-friendly coal gasification projects by China in the last decade • Inability of gasification projects to solve many fundamental problems, leading to emergency shutdown of many projects • However, reduced subsidence and damage to hydrosphere/lithosphere damage can be of help • Water consumption per tonne of coal is around 0.25 t
Energy efficiency	• Comprehensive efficiency of 50% for UCG for power production and 63% for chemical industry, as compared to only 23-30% for pulverized coal plants • Mining recovery efficiency for gasification plants of 90%, as compared to only 66% for coal mining
Economic feasibility	• Gasification of 150×10^4 tonne per annum brings advantage of economies-of-scale • Payback period of 7-12 years depending on scale of project

A major issue in UCG projects is groundwater damage, which has been studied for some time now. As back as in 1978, Humenick and Mattox [7] studied the contaminants released into groundwater due to gasification of lignite. They determined that groundwater pollutants arose from ash leachate, which contained sulfate, calcium, and hydroxide while condensed vapours contributed organics and ammonia. The lignite and clay strata were found to sorb the groundwater contaminants. Liu et al [8] analysed the organic pollutants and found the presence of phenols and benzene. They found that the flow of groundwater after gasification could be attributed to migration of the polluting substances and could be predicted using suitable numerical modelling techniques. Dong et al [9] developed a 3D hydrogeological model for optimizing water pumping and injection into UCG reservoirs. They found that more attention should be paid to control induced fractures in the roof. Thus, active regulations should be targeted in this area. They also established suitable water pumping and injection levels of (160 and 120 m^3 per day respectively). Similar tasks may be performed, at least on a block level to find suitable ranges for Indian UCG projects.

Underground coal gasification also must lead to long-term emission stabilization both in terms of gaseous and particulate emissions since many Indian cities such as the capital city of New Delhi, have been plagued by major air pollution. Further, in view of the Paris Climate Agreement, carbon mitigation and inclusion of clean coal technologies in the energy mix is of prime importance for decarbonisation of the electricity sector. Blinderman and Anderson [10] have studied the CO_2 emissions at the Chinchila UCG-IGCC project in Australia. They estimate that the CO_2 emissions of this plant showed a 55% reduction compared to a supercritical coal-fired power plant and a 25%

reduction compared to a gas-fired power plant. They also suggested that the turbine efficiency in such operations was 25% higher than that in natural gas combined cycle (NGCC) plants. Verma and Kumar [11] found the life-cycle greenhouse gas (GHG) emissions to be 0.91 and 18.00 kg-CO_2-eq/kg-H_2 in hydrogen production aided by UCG, with and without CCS respectively. It has also been estimated that the life-cycle CO_2 emissions through UCG process are only 98%, 80% and 72% of the said emissions from IGCC plants, supercritical PC plants and subcritical PC plants respectively [12]. The direction of life-cycle CO_2 emissions is showed in Figure 1. This demonstrates considerable superiority of UCG technology in terms of reduction of atmospheric CO_2 and the technology may be of use if implemented in conjunction with the clean development mechanism (CDM) or other proper carbon pricing incentives. An environmental life cycle-assessment (LCA) may be taken up for UCG in India to determine with maximum impact to the environment. Further, an understanding of the social perspectives on UCG for various segments of people, policy-makers, industry personnel, researchers and common people may be useful. This will also help in panel weighting for LCA.

Cost-effectiveness of clean coal technologies vis-à-vis renewable technologies is a key aspect that will determine their success. Olateju and Kumar [13] projected that the price of hydrogen production without CCS would be US$ 1.78/kg-H_2, which will increase to US$ 2.11-2.7/kg-H_2 with addition of CO_2 capture. The range in the latter case is because of the difference in the storage costs. Naturally, if UCG and ECBM can be implemented together in a coal rich area, it will lead to multiple benefits. This may be possible in western India since Singh and Mohanty [14] have estimated total storage potential of 3 Bt-CO_2 for Cambay and Barmer-Sanchor basins. These basins also have rich prospects for UCG in many coal mines such as Vastan, Valia, Umarsar and Kapurdi [15, 16]. Friedmann et al [17] project that UCG may produce gas at costs ranging from one-fourth to half, as compared to surface gasifiers. Sheng et al [18] estimate that UCG in conjunction with CCGT and CCS technologies will be able to produce electricity at less than €80/MWh. This is considerably cheaper than PC-CCS plants, as well as CSP plants [19, 20].

3. Perspectives on suitable acts and rules

There are two distinct regimes of minerals related regulatory regimes in India. One regime administers all minerals but petroleum and the other regime addresses petroleum itself. In some aspects, UCG is more like mineral such as coal and in some important aspect it is more like gas (such as CBM). The policy and regulatory theme for UCG has to draw mostly from regulations of coal having greater accuracy of resource determination than CBM where resource estimation has lesser accuracy. But since the resultant product is essentially gas and as such measurement, marketing and pricing may draw upon the regulatory construct of CBM.

The higher echelon of regulatory framework i.e. Mineral Act and Rules would require to be supplemented as in-situ chemical process is envisioned in the case of UCG. Generally, occurrence of minerals is defined in a manner which does not recognize in-situ chemical transition. It may appear to be a significant departure from the generally perceived notion of mineral, but then this form of utilization of coal can be suitably incorporated by defining the process of UCG as part of mineral processing which occurs in the place of occurrence with the specific reference to coal. Once this supplement is in place in the relevant Rule/ Act, issues such as lease for UCG testing and commercial production can be administered in similar lines as in the case of coal.

Current regulations permit right to multiple forms of same mineral. Petroleum and Natural Gas Rules allow exploitation of multiple forms of petroleum such as conventional and unconventional hydrocarbons. It is only that each of these forms may be regulated in isolation and also in combination. The choice really depends on the nature of resource and technology of exploitation. Concurrent development of coal mining or CBM may not be technically possible along with UCG. Co-development, therefore, is ruled out. In the nutshell, it means that Government may allow UCG while precluding the Contractor from other coal related resource or any other mineral available at different stratum.

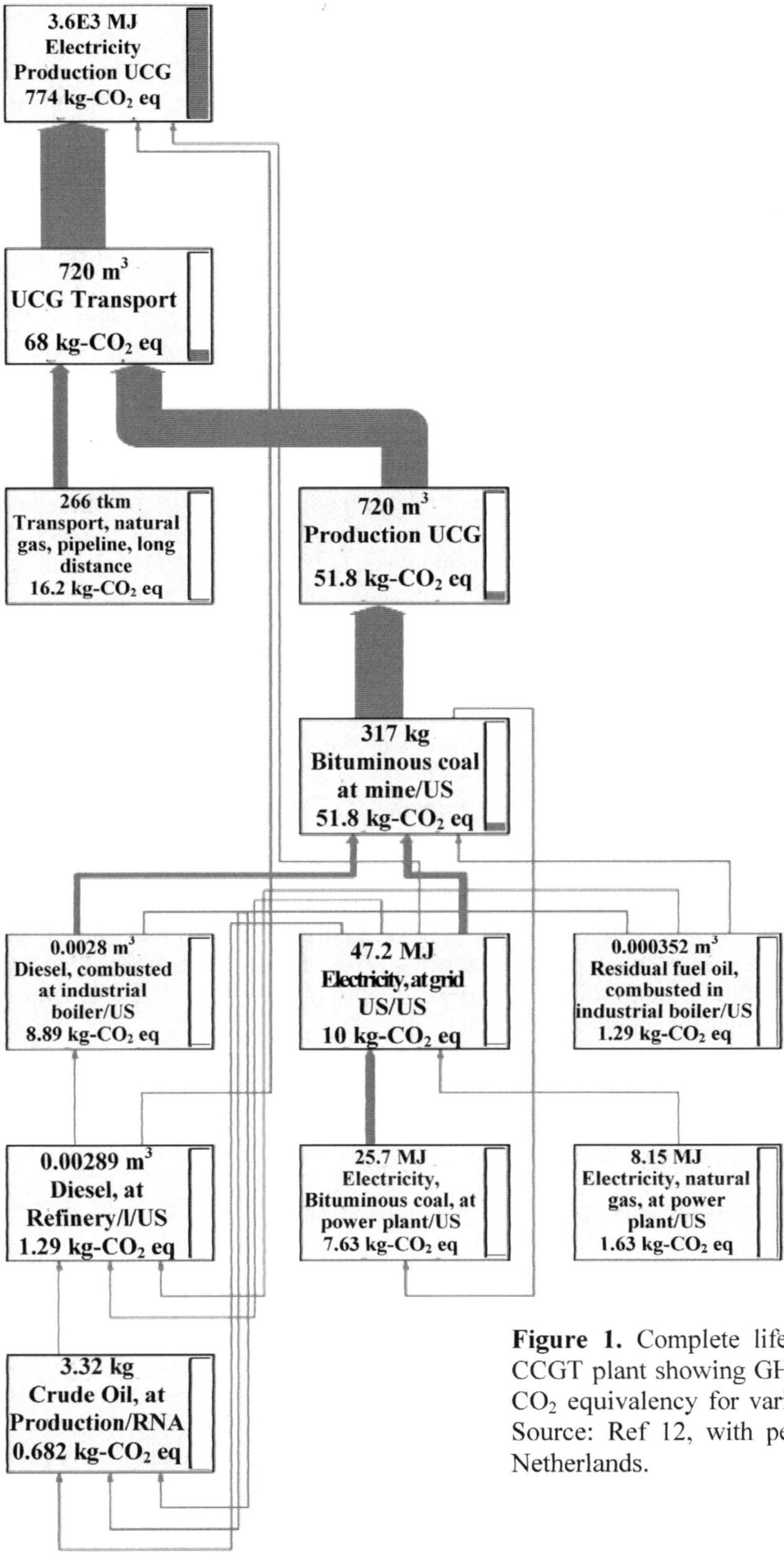

Figure 1. Complete life cycle model of UCG-CCGT plant showing GHG emissions in terms of CO_2 equivalency for various stages of life cycle. Source: Ref 12, with permission from Springer Netherlands.

IOP Conf. Series: Earth and Environmental Science **76** (2017) 012012 doi:10.1088/1755-1315/76/1/012012

4. Modalities of UCG contract development

UCG has a unique paradigm not encountered with any other resource. The development of UCG is a coupled concept as far as exploitation of resource and its utilization are concerned. The first i.e. exploitation of resource is a subject matter of mineral management and that of relevant rules and Act in that regard. The second part of the proposition i.e. utilization of UCG is subject matter beyond the scope of mineral management. Since these two parts of UCG is inseparable, the regulatory framework, therefore, needs to undertake a commercial exercise with the aim to (a) protect the value of mineral at one hand and (b) ensure commercial viability of utility project on the other. The twin requirement, though work against each other, needs to be worked upon while framing a policy or constructing a Contract.

It is evident that a minimum syngas price is assured so that exploitation is profitable and a minimum utility price (such as price of power generated) is also assured so that end use process is also profitable. Still another way to look at the utility process would be to consider it as a mere extension of in-situ processing on the surface by the same Contractor. This later consideration appears to be more realistic in the sense that it simplifies construction of a UCG Contract a lot easier. First of all, this setup resolves the conflicting aspect of two coupled processes which otherwise is bound to give rise to frequent conflicts. Only minimum syngas price is to be determined which will vary with product (such as power) price with a datum. It may be emphasized that a failed utility venture or failed UCG process in two different hands shall create a situation where public resource shall be at the mercy of contractors due to the legality involved. The provisions in the Contract should specially address situation where the process is discontinued earlier than envisaged.

5. Other specific considerations

Environment protection is a prime consideration while drafting a UCG contract. The regulation through Environment Impact Assessment (EIA) should address issues such as measurement of harmful products & mitigating provisions, Ground water preservation requirements and site restoration obligation. Various approvals would be required to be obtained from State and Central water authorities (SGWA, CGWA), MoE, National Board for Wildlife, Sanctuary, National Park, monuments etc. A body like DGMS would require to be empowered to oversee Safety and Site Restoration Activity.

In order to ensure proper control on revenue, statutory levies and profit, provisions would be required to be made to define measurement point (delivery point), unit of measurement and Standard Operating Procedure. There should be explicit provisions codifying elements of inspection, calibration, and periodicity and reporting requirements.

The contract must be designed to control the process of exploitation and utility in a controlled fashion with adequate flexibility. Generally, the control of process can be achieved by specifying timeline and corresponding extension clauses and preventive application of liquidated damage wherever required. Among other elements, these provisions may address issues relating to start time, period of different phases, submission of bank guarantee, submission and approval of development plan & utility facilities and start of commercial phase. Special considerations may be elaborated in the contract to ensure that production of syngas is synchronized with the commissioning of utility facility.

6. Conclusions

Development of UCG in India can bring several advantages in terms of efficiency, environment and economics (3E). International studies have demonstrated that UCG shows considerable advantage in terms of resource recovery as well as power plant efficiency. Further, the CO_2 emissions of UCG plants can be 70% as compared to subcritical power plants as stated in Section 2. Further, the cost of electricity (even with CCS) can be very less, when compared to combustion power plants. Having said that (in view of international experience), several apprehensions have been cast in respect to UCG in India such as fears of groundwater contamination, subsidence etc. Needless to say, to dispel off such fears, there is a need for robust and all-inclusive policy and regulatory framework, such that

IOP Conf. Series: Earth and Environmental Science **76** (2017) 012012 doi:10.1088/1755-1315/76/1/012012

sustainable resource recovery can take place to give rise to "affordable electricity to all". The outline of such a framework is shown in Figure 2.

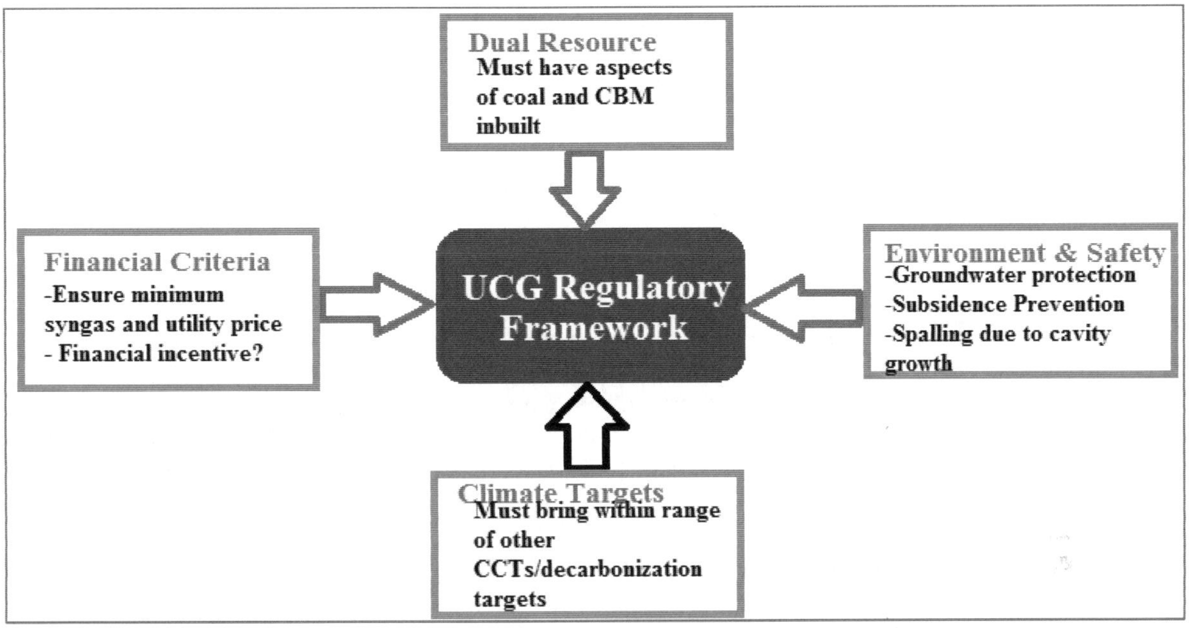

Figure 2. Proposed outline of UCG Regulatory Framework

Some of the specific regulatory recommendations in this study are:

- The regulatory framework for UCG has to take several aspects from coal having greater accuracy of resource determination than. But since the product is essentially gas, the pricing may have features from CBM pricing. Further, the Government may allow UCG whereas impeding the Contractor from additional coal related reserve or any other mineral accessible at different stratum.
- The regulatory outline is required to assume a viable implementation with the objective to safeguard the worth of mineral and guarantee commercial feasibility of utility development.
- Environmental protection should be an important part of the norms. Also, as in the case of coal mines, a separate statute needs to be developed for safety in the commercial UCG projects.

References

[1] MoC 2011 *Annual Report 2010-11* (New Delhi: Ministry of Coal, Government of India) available at http://coal.nic.in/content/annual-report-2010-11
[2] Singh U and Singh G 2016 *Strategic Planning for Energy and the Environment* **36** 43-66
[3] Iyengar R K and Haque R 1991 *Fuel Processing Technology* **27** 247-262
[4] Krishnudu T, Madhusudhan B, Reddy S N, Rao K S and Vaidyeswaran R 1989 *Fuel Processing Technology* **23** 233-256
[5] Shahi R V 2010 *CO_2 Sequestration Technologies for Clean Energy* ed S Z Qasim and M Goel (New Delhi: Daya Publishing Press) p 13-22
[6] Mao F 2016 *Natural Gas Industry B* **3** 312-322
[7] Humenick M J and Mattox C F 1978 *Water Research* **12** 463-469
[8] Liu S, Li J and Dong D 2007 *Journal of China University of Mining and Technology* **17** 467-

IOP Conf. Series: Earth and Environmental Science **76** (2017) 012012 doi:10.1088/1755-1315/76/1/012012

472
[9] Dong D, Zhao M, Lin G, Zhao W, Zhang K and Li N 2016 *Mine Water and the Environment* **35** 398-404
[10] Blinderman M S and Anderson B 2004 *ASME 2004 Power Conference* (Baltimore) p 473-479
[11] Verma A and Kumar A 2015 *Applied Energy* **147** 556-568
[12] Hyder Z, Ripepi N S and Karmis M E 2016 *Mitigation and Adaptation Strategies for Global Change* **21** 515-546
[13] Olateju B and Kumar A 2013 *Applied Energy* **111** 428-440
[14] Singh A K and Mohanty D 2015 *Carbon Capture, Storage and, Utilization: A possible climate change solution for energy industry* ed M Goel, M Sudhakar and R V Shahi (New Delhi: TERI Press) p 133-147
[15] Akanksha, Singh A K, Mohanty D and Jena H M 2017 *Nexgen Technologies for Mining and Fuel Industries* (New Delhi:Allied Publishers) p 1425-1430
[16] Sanga B, Mohanty D, Singh A K and Singh R K 2017 *Nexgen Technologies for Mining and Fuel Industries* (New Delhi:Allied Publishers) p 1431-1436
[17] Friedmann S J, Upadhye R and Kong F 2009 *Energy Procedia* **1** 4551-4557
[18] Sheng Y., Benderev A, Bukolska D, Eshiet K I, Gama C D, Gorka T, Green M, Hristov N, Katsimpardi I, Kempka T and Kortenski J 2016 *Mitigation and Adaptation Strategies for Global Change* **21** 595-627
[19] Singh U and Rao A B 2016 *Energy Procedia* **90** 326-335
[20] Krishnamurthy P, Mishra S and Banerjee R 2012 *Energy Policy* **48** 407-419

Institute of Physics
Dirac House, Temple Back
Bristol BS1 6BE UK

ISSN: 1755-1307
ISBN 978-1-5108-4509-1